# SPATIAL INTERACTION MODELS: FORMULATIONS AND APPLICATIONS

# STUDIES IN OPERATIONAL REGIONAL SCIENCE

Folmer, H., Regional Economic Policy. 1986. ISBN 90-247-3308-1.

Brouwer, F., Integrated Environmental Modelling: Design and Tools. 1987. ISBN 90-247-3519-X.

Toyomane, N., Multiregional Input–Output Models in Long-Run Simulation. 1988. ISBN 90-247-3679-X.

Anselin, L., Spatial Econometrics: Methods and Models. 1988. ISBN 90-247-3735-4.

Fotheringham, A.S. and O'Kelly, M.E., Spatial Interaction Models: Formulations and Applications. 1989. ISBN 0-7923-0021-1.

# Spatial Interaction Models: Formulations and Applications

by

## A.S. Fotheringham
*State University of New York at Buffalo, USA*

and

## M.E. O'Kelly
*The Ohio State University, USA*

KLUWER ACADEMIC PUBLISHERS
DORDRECHT / BOSTON / LONDON

Library of Congress Cataloging in Publication Data

Fotheringham, A. Stewart.
   Spatial interaction models : formulations and applications / A.
Stewart Fotheringham, Morton E. O'Kelly.
     p.   cm. -- (Studies in operational regional science)
   Bibliography: p.
   Includes index.
   ISBN 0-7923-0021-1 (U.S.)
   1. Space and economics--Mathematical models.  2. Geography,
Economic--Mathematical models.  3. Industry--Location--Mathematical
models.   I. O'Kelly, Morton E., 1955-   . II. Title. III. Series.
HB199.F65 1988
338.6'042'0724--dc19                                      88-29416
                                                             CIP

ISBN 0-7923-0021-1

Published by Kluwer Academic Publishers,
P.O. Box 17, 3300 AA Dordrecht, The Netherlands.

Kluwer Academic Publishers incorporates
the publishing programmes of
D. Reidel, Martinus Nijhoff, Dr W. Junk and MTP Press.

Sold and distributed in the U.S.A. and Canada
by Kluwer Academic Publishers,
101 Philip Drive, Norwell, MA 02061, U.S.A.

In all other countries, sold and distributed
by Kluwer Academic Publishers Group,
P.O. Box 322, 3300 AH Dordrecht, The Netherlands.

Printed in The Netherlands

*To Iain and Matthew*

## Table of Contents

# List of Tables

# List of Figures

xvi

# FOREWORD

This book is a response to a perceived omission in the comprehensive understanding of one of the most important topics within regional science and analytical human geography. While vast amounts of research into spatial interaction modelling exist, most notably in journals such as *Environment and Planning A, Geographical Analysis* and the *Journal of Regional Science*, there has yet to be an advanced discussion of the general concepts involved in the widespread application of such models. This book attempts to fill this gap and so is intended to appeal not only to academic spatial analysts but also to practitioners in fields such as marketing, demography, and transportation planning.

The book is in two sections. Section 1, consisting of four chapters, is devoted to a description and understanding of some general issues involved in designing, formulating, and calibrating spatial interaction models. Section 2, consisting of five chapters, contains specific examples of the use and application of spatial interaction models in such diverse areas as locational analysis, demography, network design, retailing, and urban transportation analysis.

In Chapter 1 we define spatial interaction and discuss its importance as a link between specialisation and economic development. We introduce some of the basic ideas in spatial interaction modelling and comment on the measurement of variables and their most appropriate functional forms. Chapter 2 contains an outline of a series of spatial interaction models, each derived in a consistent manner through information minimising techniques. Numerical examples are used to illustrate the key features in the design of such models and also the variations that can be built into this design. A relatively new concept in spatial interaction modelling, that of 'relaxed' interaction models, is discussed. This chapter provides a notational scheme and a set of model development tools which are subsequently put to use in the rest of the book.

Chapter 3 demonstrates the calibration of spatial interaction models by both maximum-likelihood and by least squares regression. Since the large majority of spatial interaction models are non-linear, a set of appropriate linear transforms for the latter calibration technique is described. Numerical examples of both types of calibration procedure are given for a family of models. Also discussed here are computer algorithms for spatial interaction model calibration and appropriate diagnostics for this procedure. Finally, some of the problems that can be encountered in model calibration are described and some solutions to these problems are outlined. In the fourth and final chapter of this part of the book, the relationship between micro and macro models of spatial interaction models is explored. The former focus on the behaviour of the individual in terms of making spatial choices, while the latter focus on aggregate flows between places. The opportunity is taken here to review and extend the debate concerning the context-dependency of spatial interaction models.

Part 2 begins in Chapter 5 with a demonstration of the applicability of spatial interaction models in demographic analysis. The application of these models as an aid to understanding the determinants of broad-scale movements of populations within countries is examined and a detailed empirical example of migration within the United Kingdom is presented, In Chapter 6 the importance of market share analysis in retailing and the importance of spatial interaction modelling in predicting market shares of retail outlets are described. Topics considered include the relative attributes of stores that influence a consumer's store selection; the effect of various types of prejudice on store selection and how these can be modelled; and the influence of personal characteristics on consumer spatial behaviour. We describe a typical market share analysis of supermarket shopping that can be applied to shopping patterns in any city. This type of analysis is useful for determining the optimal location of a new outlet; determining the optimal store in a chain in which to expand operations; deriving realistic retail targets for individual stores; determining how the actions of competitors are likely to affect a particular store; and for producing potential revenue surfaces.

In Chapter 7 the application of spatial interaction models to forecast complex trip patterns within cities is explored. In particular, the role of multi-purpose, multi-stop and multi--destination travel in urban areas and the consequences of such behaviour for the spatial arrangement of various types of land uses within urban areas are examined. The design and analysis of a travel diary survey is reported.

Spatial interaction models and location-allocation models represent two powerful techniques available to the spatial analyst interested in the location of activities. Chapter 8 reviews the connections between these two types of models and formulates a consistent optimisation framework to represent an interaction-based, location-allocation model. A solution procedure for this model is then presented. The model is applied to a problem of siting facilities in 20 of a possible 181 locations under several scenarios. The second part of this chapter suggests the application of the spatial interaction based location-allocation model to the design of a two-level emergency service.

The final chapter of the book answers the question "Where should facilities such as air terminals and express delivery sorting centers be located so as to connect cities in a network efficiently?" The problem has contemporary relevance because several air carriers, especially in the United States, operate a highly simplified, sparse network organised around a 'hub and spoke' system. Hubs are a type of facility located in the network to provide a switching point for flows between interacting nodes. Siting a hub poses a problem for the spatial interaction analyst because the indirect routing of flows through central facilities implies a distortion of conventional transportation costs, which may require a re-evaluation of the demand for interaction between any two nodes. We attempt to add to the growing literature on this topic by examining three situations: (i) the siting of a single facility to service the interactions between a fixed set of nodes; (ii) the siting of two or more planar facilities to service the interactions; and (iii) the siting of nodes at discrete positions in a network to serve as switching points. Models for each situation are illustrated using intercity airline passenger interactions within the United States.

Given the complexity of some of the above topics, it is not surprising that part of the discussion within this text will be best understood and explained using quantitative and analytical methods. The aim throughout is to concentrate on a clear exposition of the widespread applicability of the concepts that are included under the banner of spatial interaction modelling. To this end, we have made an attempt to clarify the more complex issues with simple numerical examples and, where possible, with actual empirical applications. We thus hope to encourage the reader to explore what we consider to be a useful and important topic, although some may find that an introductory text such as that provided by Haynes and Fotheringham (1984) is a useful precursor to this book.

# Acknowledgements

The first author thanks the trustees of the Leverhulme Foundation in the United Kingdom for the award of a Leverhulme Fellowship during the 1987/88 academic year which provided very valuable research time and which enabled the book to be completed on schedule. He also thanks the members of the department of Town Planning at the University of Wales Institute of Science and Technology for providing generous research facilities and assistance during the tenure of the fellowship. Special thanks to Mike Batty, Neil Wrigley and Paul Longley.

The second author thanks the National Science Foundation for supporting his research on the topics of spatial interaction and locational analysis (NSF #SES-8309590 and #SES-8644369). He also thanks the department of Geography at The Ohio State University for facilities and computer equipment which were essential to the completion of the manuscript. Special thanks to John Rayner, Duane Marble and Allaire Shaw.

Thanks also to Barbara Fotheringham who stoically and efficiently typed sections of the book while coping with the latter stages of pregnancy. Susan O'Kelly helped with proof reading.

Specific acknowledgements for permission to use published material are as follows:

In **Chapter 2**, The Geographical Society of Ireland and J.A. Walsh for data used in Tables 2.5, 2.6 and Figure 2.2. The *Annals of the Association of American Geographers* for Table 2.7. *Environment and Planning A* for Table 2.8. In **Chapter 3**, *Environment and Planning A* for Tables 3.1 and 3.2. In **Chapter 5**, *Journal of Regional Science* for Table 5.1; *Environment and Planning A* for Figure 5.2 and Table 5.2. Robin Flowerdew for providing the UK migration data and for assisting with its analysis. Henk Scholten of the Central Bureau of Statistics in The Netherlands for providing the Dutch migration data. In **Chapter 6**, *Environment and Planning A* for Figure 6.2. Parts of **Chapter 7** benefitted from discussions with Eric Miller. In **Chapter 8**, van Nostrand for Tables 8.1 and 8.2 and Figures 8.1 and 8.2. Also *Regional Studies* for Tables 8.3, 8.4 and 8.5, and Figures 8.3-8.8. Parts of Chapter 8 are the result of joint research with Jim Storbeck. In **Chapter 9**, Tables 9.1 and 9.2, and Figures 9.3, 9.4 and 9.5 are from *Transportation Science*. Tables 9.3 and 9.4, and Figure 9.6 are from the *European Journal of Operational Research*.

# PART 1

FORMULATIONS

# CHAPTER 1

## THE ELEMENTS OF SPATIAL INTERACTION MODELLING

### 1.1 Introduction

The criticism is sometimes levelled at those who attempt to analyse human spatial behaviour that such behaviour is intrinsically unpredictable. This criticism can be countered by pointing out that certain types of spatial behaviour can be predicted quite easily. For example, nobody resident in Cardiff, Wales will travel exclusively to purchase tomorrow's groceries in London, England, some 160 miles away. The banality of this prediction only serves to underline the widespread recognition of the fact that individuals are constrained in certain actions by space. Generally, however, this constraint is not nearly so clear as described above; while individuals do have a tendency to minimise the distance they travel to partake in some activity such as shopping, longer distances than are absolutely necessary are often travelled in order to take advantage of destinations offering greater opportunities. Thus, the residents of Cardiff, while not travelling to London for grocery shopping, may well bypass more local stores to patronise a more distant superstore. It is understanding this trade-off between the constraint of distance and the attraction of increased opportunities which is at the heart of spatial interaction modelling. Before exploring this more thoroughly, it is useful to define in broad terms what is meant by spatial interaction and what are the purposes of spatial interaction modelling.

Spatial interaction can be broadly defined as movement or communication over space that results from a decision process. The term thus encompasses such diverse behaviour as migration, shopping, travel-to-work, the choice of health-care services, recreation, the movement of goods, telephone calls, the choice of a university by students, airline passenger traffic, and even attendance at events such as conferences, theatre and football matches. In each case, an individual trades off in some manner the benefit of the interaction (the purchase of goods at a store, for example) with the costs that are necessary in overcoming the spatial separation between the individual and his/her possible destination. Thus, while no person from Cardiff would consider undertaking grocery shopping in London, they may consider occasionally travelling to London for non-grocery shopping. The friction of distance, or the 'distance-decay', associated with the former type of activity is evidently much greater than that associated with the latter. It is the pervasiveness of this type of trade-off in spatial behaviour that has made spatial interaction modelling so important and the subject of such intense investigation. As Olsson (1970, p.233) states:

> The concept of spatial interaction is central for everyone concerned with theoretical geography and regional science....Under the umbrella of spatial interaction and distance decay, it has been possible to accommodate most model work in transportation, migration, commuting, and diffusion, as well as significant aspects of location theory.

### 1.2 Spatial Interaction and Economic Development

Spatial interaction is inexorably linked with economic development through the process of specialisation as described in Figure 1.1. The very early stages of a system's economic development are characterised by high levels of self-sufficiency and, consequently, comparatively low levels of spatial interaction. Economic development takes place by increased specialisation which allows increased efficiency and production. A necessity of specialisation, however, is trade and communication. Thus, while economic development is determined by specialisation, specialisation is only possible through spatial interaction.

Figure 1.1  **The Link Between Spatial Interaction and Economic Development**

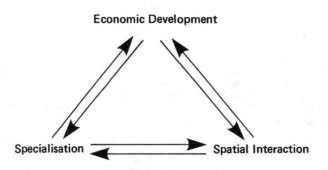

The prosperity enjoyed in advanced economic societies is hence only possible through the maintenance of high levels of spatial interaction. This dependency is the reason why the development of infrastructure to aid spatial interaction is so prominent in our environment. We are surrounded with the evidence of this in the form of roads, bridges, railways, canals, airports, telephones, mail boxes and post offices, and computer networks. Movement over space plays an important role in human activities and decisions that affect such movement are often some of the most major in our lives (such as the decision to migrate to another city or country) or occur very frequently (such as the decision about which shops to patronise).

1.3  The Purposes of Spatial Interaction Modelling

All mathematical modelling can have two major, sometimes contradictory, aims: explanation and prediction. In terms of spatial interaction models, explanation involves determining through model calibration the attributes of locations that promote flows of people, goods or ideas between them. Used in this mode, the key feature of an interaction model is that it be properly specified in a detailed manner so that the effect of each determinant of interaction can be assessed though an associated parameter estimate. Three separate types of explanatory spatial interaction models exist: those that yield insight into interaction patterns by providing information on the attributes of both the origins and the destinations of the interactions; those that provide information only on destination characteristics; and those that provide information only on origin characteristics. The first type of models are termed unconstrained, or total flow constrained; the second type are known as production-constrained; and the third type are known as attraction-constrained.

A fourth type of spatial interaction model exists, the doubly constrained or production-attraction-constrained model. Its major purpose is predictive rather than explanatory in that it takes the propulsiveness of origins and the attractiveness of destinations as given exogenously and merely seeks to allocate a known number of outflows and inflows to links between these origins and destinations. It provides no information on what characteristics make a destination attractive or what characteristics make an origin unattractive but it generally provides high levels of predictive or replicative power within these constraints.

While a detailed discussion of each of these types of spatial interaction models is left to Chapter 2, we provide here examples of circumstances in which each of the above models would be most usefully applied. Consider a spatial system in which migration is taking place between a set of origins and destinations. Several questions might be asked of this system:

(i) Why are there large flows between some places but only small flows between others? What characteristics of places promote large flows between them? To answer this question, an unconstrained model would probably be most appropriate.

(ii) What are the characteristics of destinations that make them attractive for migrants? Here a production-constrained model, which takes as given the number of flows from each origin and allocates them to destinations, is most appropriate.

(iii) What are the characteristics of origins that make them unattractive to individuals and so produce a large number of outmigrants? In this case, an attraction-constrained model, which takes as given the number of interactions into each destination and allocates these interactions to a set of origins, is most appropriate.

(iv) Given that the outflow from each origin and the inflow into each destination are known, what is the pattern of flows likely to be between these nodes?  Here, a doubly-constrained model is most appropriate.

In reading the above set of scenarios, a newcomer to spatial interaction modelling might be forgiven for asking the question "Since the unconstrained model provides the most information, why not use it to answer all of the questions regarding spatial interaction?". Unfortunately, as depicted in Figure 1.2, there is a trade-off between the quality and quantity of information provided by different types of spatial interaction models.  Ideally, we want a model in the top left-hand corner of the diagram; a model providing large amounts of high-quality information.  Our experience tells us that the closest we can come to this ideal is with the use of one of the two singly constrained models, the production-constrained and the attraction-constrained models.   Consequently, in the subsequent chapters of this book, much emphasis is placed on the use of such models. The unconstrained model provides a large quantity of information although its quality is generally not considered to be at an acceptable level, whereas the doubly constrained model provides high quality information but in generally unacceptably low quantities.  We will demonstrate how information is obtained from spatial interaction models through model calibration in a subsequent chapter.

Figure 1.2  **Trade-off Between the Quantity and Quality of Information Provided by Spatial Interaction Models**

QUANTITY

| | Large | Small |
|---|---|---|
| | | |

| | | |
|---|---|---|
| High | production-constrained /attraction-constrained | production -attraction constrained |

QUALITY

| | | |
|---|---|---|
| Low | unconstrained | |

So far, we have discussed in general terms the uses of spatial interaction models. We now turn to a list of more specific questions that can be answered through spatial interaction analysis. This list is not exhaustive but is sufficiently comprehensive to give the reader a sense of whether the techniques described in the next few chapters are relevant to his/her needs.

(i) Is there sufficient demand to establish a profitable air service between two cities? What frequency of air service will be sufficient to meet the expected demand?

(ii) Is the investment in a new rail service, road system or bridge cost-effective given the expected demand?

(iii) What will be the effect on traffic patterns of building a new superstore?

(iv) What will be the effect on travel patterns of building a tunnel under the English Channel between England and France?

(v) What increase in housing demand can be expected in particular parts of a city due to the development of a new industrial center?

(vi) Where is the optimal location for a new supermarket?

(vii) What is the expected turnover at a store?

(viii) How sensitive are various types of movements to changes in travel costs?

(ix) What effects do boundaries (actual and perceived) have on interaction patterns?

(x) What differences are there in travel behaviour between different types of individuals?

(xi) Are we becoming less constrained by space over time?

(xii) What effects do changes in unemployment levels have on migration patterns?

(xiii) What is the most efficient arrangement of hubs in an airline hub and spoke network?

While some of these questions will be addressed explicitly in the main body of the book, others are posed here as examples of puzzles for which we supply formulation guidelines and solution methods.

## 1.4  The Basics of Spatial Interaction Modelling

The starting point of nearly all spatial interaction analysis is an origin-destination matrix of the form described in Figure 1.3 in which there is a set of flows, $T$, between m origins and n destinations.  We also have a compatible m x n matrix $C$, whose elements depict the spatial separation between origins and destinations (generally in terms of distance, cost, or time); an m x p matrix of origin propulsiveness measures, $V$; and an n x q matrix of destination attractiveness variables, $W$.  The symbols p and q denote the number of propulsiveness and attractiveness variables, respectively.

## Figure 1.3  The Basic Elements of Spatial Interaction Modelling

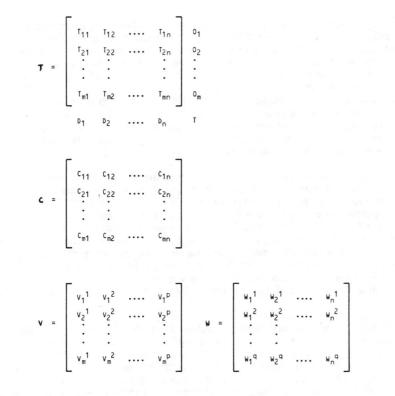

By summing the observed interaction matrix across each row, we obtain the observed outflow from each origin, $O_i$, and by summing the observed interaction matrix down each column we obtain the observed inflow into each destination, $D_j$.  The total interaction in the system is represented by T.   Clearly,

$$T = \Sigma_j D_j = \Sigma_j \Sigma_i T_{ij} \tag{1.1}$$

or, equivalently,

$$T = \Sigma_i O_i = \Sigma_i \Sigma_j T_{ij} \tag{1.2}$$

In essence, the task of the spatial interaction modeller is to relate the values in the matrix **T** to the corresponding values in the matrices **V**, **W** and **C**.   In a production-constrained model, where the investigation of the origin propulsiveness variables is ignored, the matrix **V** is replaced by an m x 1 matrix, **O**, whose elements consist of the row totals of the matrix **T**.   Similarly, in an attraction-constrained model, where the investigation of the destination attractiveness variables is ignored, the matrix **W** is replaced by a 1 x n matrix, **D**, whose elements consist of the column totals of the matrix **T**.   In doubly constrained models, both the origin and destination matrices are replaced in this manner.   Consequently, in the calibration of such a model the only explanatory information that is obtained concerns the role of spatial separation in determining interaction patterns.   We now turn to a brief discussion of the types of variables that might be encountered in the matrices **V**, **W** and **C**.

The matrix **V** contains variables that determine the propulsiveness of an origin; that is, variables influencing the volume of flows from each origin.   These variables will clearly vary with the context of the interaction taking place.   The factors that prompt individuals to migrate are not the same as those inducing people to travel to work or to spend money in shops.   Listed in Table 1.1 are some of the typical variables that might be included in the matrix **V** for two types of spatial interaction - shopping expenditure flows and migration.   Also included are the anticipated relationships between each variable and flow volumes.   For instance, in the case of migration, the expected relationship between $v_i^4$ and $O_i$ is positive indicating that as the crime rate at an origin increases, it can be expected that more people will want to move away from that area.   Conversely, as wage rates increase, fewer people will want to leave.

In a similar manner, the matrix **W** contains variables that determine the attractiveness of a destination, or the $D_j$ values.   Again, exactly what variables are relevant depends on the particular interaction system being investigated.   Typical attributes of destinations that might be important in examining shopping expenditure flows and migration patterns are listed in Table 1.2.   Notice that in the shopping example, the attributes of the origins and destinations that might be included in a spatial interaction model of shopping expenditures are very different because the origins represent the amount of money available for shopping in various residential neighbourhoods while the destinations are particular shops or shopping centers.   Conversely, in the migration example the two sets of attributes are virtually identical because the origins and destinations are the same entities - most usually, cities or regions.

The significance of a variable in determining interaction patterns is obtained through model calibration; a topic covered in Chapter 3. It is sufficient to say at this stage that a vast amount of empirical research has been undertaken in attempting to identify the relevant attributes of origins and destinations for particular types of interaction. More detail is given on such studies in later chapters, particularly in the discussions of the application of spatial interaction models to retailing and migration. However, there appears to be little consistency across studies and it remains, to some degree, a matter of experimentation as to which variables will be useful in explaining a particular set of interactions. It is also an interesting research question as to why some variables appear to be relevant in certain situations but not in others.

Table 1.1 **Examples of Origin Attributes for Two Types of Spatial Interaction**

**SHOPPING EXPENDITURES**

| Attribute of Residential Area ($v_i$) | Relationship with $O_i$ |
|---|---|
| 1  Average household income | + |
| 2  Number of households | + |
| 3  Average family size | + |

**MIGRATION**

| Attribute of Origin ($v_i$) | Relationship with $O_i$ |
|---|---|
| 1  Number of people | + |
| 2  Cost of living | + |
| 3  Level of amenities | - |
| 4  Crime rate | + |
| 5  Wage rate | - |
| 6  Unemployment rate | + |

Table 1.2 **Examples of Destination Attributes for Two Types of Spatial Interaction**

**SHOPPING EXPENDITURES**

| Attribute of Store ($w_j$) | Relationship with $D_j$ |
|---|---|
| 1  Variety of goods | + |
| 2  Ease of returning goods | + |
| 3  Average prices | - |
| 4  Degree of crowding | - |
| 5  Number of parking spaces | + |
| 6  Quality of goods | + |

Table 1.2 (CONTINUED)

---

**MIGRATION**

---

| Attribute of Destination ($w_j$) | Relationship with $D_j$ |
|---|:---:|
| 1  Number of people | + |
| 2  Cost of living | - |
| 3  Level of amenities | + |
| 4  Crime rate | - |
| 5  Wage rate | + |
| 6  Unemployment rate | - |
| 7  Average house price | - |
| 8  Average temperature | + |

---

The definition of what constitutes a relevant measure of spatial separation in the matrix C is also problematic and has also been the subject of fairly intense experimentation. Three general measures dominate the literature: these being distance, travel cost and travel time. Each of these variables can be measured either objectively, by direct observation, or subjectively, by questioning individuals about their cognition of spatial separation. Cadwallader (1975), for example, provides a demonstration of an interaction model, albeit a very simple one, calibrated with cognitive data rather than objective data. However, while the idea of using cognitive data is interesting, it has not found widespread application because it is an extremely time-consuming and data-intensive procedure. Consequently, in the interests of remaining relatively pragmatic, we will not pursue this particular line of research here and instead will focus on the more usual procedure of obtaining objective measures of spatial separation.

Of the three measures described above, distance is by far the most commonly used. In studying migration, for example, it makes little sense to think that travel cost or travel time will have any great effect on migration patterns. This is because the primary role of spatial separation in migration is not to measure the cost of overcoming space but rather is to measure the impact of distance on information decay about places (Schwartz, 1973). Migrants tend to minimise the uncertainty about a move by favouring closer destinations over more distant ones. Even in the analysis of intraurban interactions such as shopping trips or journey-to-work, physical distance is often used as a surrogate for travel cost or travel time because of the problems of measuring the last two variables. In what follows we will generally assume that for intraurban interactions spatial interaction is measured by travel <u>cost</u> while for interurban interactions spatial separation is measured by <u>distance</u>.

Using distance rather than cost to measure spatial separation in systems with many nodes has the further advantage that a large number of distances can be obtained with minimal effort using the coordinates of points in the following formula:

$$d_{ij} = k \ [|x_i - x_j|^p + |y_i - y_j|^p]^{\ 1/p} \tag{1.3}$$

where the coordinates of origins and destinations are represented by $(x_i, y_i)$ and $(x_j, y_j)$, respectively. The notation $d_{ij}$ is used to represent the distance between i and j. The parameter k is a scale conversion factor, while p can be calibrated to give a good correspondence between actual travel distances and the modelled values. Typically p ranges between 1 and 2 but for intraurban distances p is generally taken as 1.0 to represent rectangular distances; the value for interurban distances is generally taken as 2.0 representing straight line distance. (When $p < 1$ the resulting $d_{ij}$ values do not obey the triangle inequality.) An interesting application of the formula in equation (1.3) was made by Love and Morris (1972; 1979); they calibrated k and p values to fit $d_{ij}$ to 15 x 15 distance tables for each of 7 different study areas. The observations are the actual shortest routes by road between the pairs of locations. The results of their technique applied to intra-urban data in Columbus Ohio, show that the 105 distances being modelled are best represented by values of $k = 1.18$ and $p = 1.47$. (A variety of other distance measures and study areas are reported but with substantially similar results.)

A problem arises, however, in the use of this formula to obtain distances when interaction data are only available between origin <u>zones</u> and destination <u>points</u>. Such might be the case, for example, when shopping data at stores are available by residential neighbourhood. While there is no problem in calculating distances between a zone centroid and points outside that particular zone using equation (1.3), a problem arises whenever intrazonal distances have to be calculated. Clearly, if a destination point were at a zone centroid the calculated distance would be zero whereas the average distance travelled from within the zone would, in reality, be positive. This is a problem caused by using aggregate data and assuming population to be located at zone centroids rather than distributed continuously across space, but nonetheless it is a problem that is frequently encountered in intraurban analyses of spatial interaction. Approximate solutions can be found in Eilon, Watson-Gandy and Christofides (1971) who provide measures of intrazonal distances for zones of varying shapes and under various assumptions about the population distribution and the point location within the zone. Based on their results, Fotheringham (1988a) has suggested the following formula for deriving rectangular distances within zones that are approximately circular:

$$d = 0.846(1.693)^{z/r} \cdot r \qquad (1.4)$$

where z is the distance between the zone centroid and the destination point and r is the radius of a circle whose area is equal to that of the zone. When the point is located at the zone centroid, $z = 0$ and $d = 0.846r$; when the point is located on the circumference of the zone, $z = r$ and $d = 1.432r$; when the point is located between the center and the circumference, $0.846r < d < 1.432r$. The equivalent formula given by Fotheringham (1988a) for zones that are approximately rectangular is:

$$d = (1/4 + z/4q)(a + b) \qquad (1.5)$$

where,

$$q = [(a/2)^2 + (b/2)^2] \qquad (1.6)$$

and where a and b represent the length and width of the rectangle.

So far, we have defined three sets of variables to explain interactions: a set of origin-propulsiveness variables; a set of destination-attractiveness variables; and a set of spatial separation measures. We briefly mention at this stage the possibility of including a fourth basic element, that of destination situation, which has been examined by Fotheringham (1983a; 1984; 1985; 1986). It is thought that the location of a destination with respect to its 'competitors' might affect the probability of an individual selecting that destination. An example of such behaviour is when an individual selects a store in close proximity to other stores and so minimises his/her anticipated travel costs beyond the first store visited. In essence, this is why shopping centers have proven so popular with consumers (O'Kelly, 1983a; 1983b). A detailed description of this particular topic is postponed until Chapter 7.

## 1.5   A General Spatial Interaction Model

From the above discussion, a general spatial interaction model can be formulated as follows:

$$T_{ij} = f(\mu_1 v_i^1, \mu_2 v_i^2, \ldots ,\mu_p v_i^p; \alpha_1 w_j^1, \alpha_2 w_j^2, \ldots ,\alpha_q w_j^q; \beta c_{ij}) \qquad (1.7)$$

where $T_{ij}$ represents a flow between i and j, $v_i$ represents a variable measuring the propulsiveness of i, $w_j$ represents a variable measuring the attractiveness of j, and $c_{ij}$ represents the spatial separation between i and j. The parameters $\mu$, $\alpha$ and $\beta$ reflect the relationship between each of these variables and $T_{ij}$. In much of what follows we will concentrate on the basic structure of the model rather than on a detailed examination of the composition of each set of variables. In these instances, we can represent the structure of the model in a simpler manner by assuming that there is only one relevant propulsiveness variable and one relevant attractiveness variable so that (1.7) can be rewritten as:

$$T_{ij} = f(\mu v_i; \alpha w_j; \beta c_{ij}) . \qquad (1.8)$$

We stress, however, that this assumption is only made to simplify what would otherwise be cumbersome notation; we do not mean to imply that complex phenomena such as spatial interaction patterns can be encapsulated in a simple, three variable model. However, the historical development of spatial interaction modelling was dominated by this type of simple model (the gravity model) and a three variable model can yield surprisingly high levels of goodness-of-fit. Reviews of the development of gravity and spatial interaction models can be found elsewhere (inter alia, Haynes and Fotheringham, 1984; Senior, 1979).

The exact functional form of each type of variable in the model is subject to varying degrees of conjecture. There is virtual unanimity of opinion that the site specific variables, the $v_i$s and the $w_j$s, are generally best represented as power functions;

$$f(\mu v_i) = v_i^\mu \qquad (1.9)$$

$$f(\alpha w_j) = w_j^\alpha . \qquad (1.10)$$

The exact form of the separation function is more in doubt although two forms dominate the literature. These are the power function,

$$f(\beta c_{ij}) = c_{ij}^\beta \qquad (1.11)$$

and the exponential function,

$$f(\beta c_{ij}) = \exp(\beta c_{ij}). \qquad (1.12)$$

Since interaction is expected to <u>decrease</u> as separation increases, all else being equal, the parameter $\beta$ is often written explicitly as a negative value so that, for example, the exponential function is often written as:

$$f(\beta c_{ij}) = \exp(-\beta c_{ij}) \qquad (1.13)$$

and in what follows we will follow this particular convention; this means that the distance parameter is discussed in terms of its magnitude or absolute value, since it is entered into (1.13) with a negative sign. Those cases where there is a genuinely positive impact of distance on interaction are sufficiently distinctive that no ambiguity in the use of our sign convention should arise.

There are four issues involved in the debate as to whether a power or exponential function of distance should be employed in spatial interaction models and we now briefly consider each of them. The first concerns parameter comparability and is only an issue if values of $\beta$ are to be compared across studies. This is a particularly useful exercise, however, and can yield important information regarding variations in behaviour over space and/or over time. In practical applications of spatial interaction models, parameter incomparability is a major issue if we want to apply a model to a system other than the one in which it has been calibrated. In such instances, it is clearly important to obtain estimates of parameters that are independent of the scale of the system and that are independent of the units in which separation is measured. It can easily be demonstrated that the power function supplies desirable scale-independent parameter estimates while the exponential function does not. Let $c_{ij}{}'$ be a measure of travel cost in pounds sterling and let $c_{ij}{}^d$ be an equivalent measure in dollars. Suppose that the exchange rate is 1 pound = k dollars. Then,

$$(c_{ij}{}')^{-\beta} = (kc_{ij}{}^d)^{-\beta} = k^{-\beta}(c_{ij}{}^d)^{-\beta} \tag{1.14}$$

so that the distance-decay parameter, $\beta$, is the same whether costs are measured in dollars or sterling. The variation in the measurement of costs is subsumed in a constant term. However, if the equivalent transformation is undertaken with the exponential function,

$$\exp(-\beta c_{ij}{}') = \exp(-\beta k c_{ij}{}^d) = \exp(-\beta^* c_{ij}{}^d) \tag{1.15}$$

a different parameter, $\beta^*$, will be estimated in the model calibration. If $k > 1$, $\beta^* > \beta$ and if $k < 1$, $\beta^* < \beta$. A similar problem of scale-dependency exists when costs are measured in the same units but in systems at different scales. Thus, a model with an exponential cost function calibrated with traffic flows from a major city could not be used to forecast traffic flows in a medium or small urban area.

The second issue in the debate over the functional form of the spatial separation variable is that raised by Evans (1970) and concerns expected cost increases. Without going into the details of the explanation, which are relatively straightforward and similar in logic to the above discussion, a summary of Evans' results is presented in Figure 1.4. Suppose in an analysis of passenger flows on public transit within a major city, costs are to be increased along certain routes. Two possible consequences of this action can result: the selective increase in costs will alter the whole trip matrix; or the trip matrix will remain stable. Also, two types of cost increase are possible; each fare can be increased by a constant multiple or a constant amount. This produces four scenarios under each of which one of the two spatial separation functions is appropriate and the other is inappropriate. For instance, if a multiplicative cost increase is to be applied and this is expected to alter the trip matrix, an exponential cost function should be employed. Conversely, if the cost increase is to be additive, a power function is more appropriate.

**Figure 1.4  Replication of Expected Travel Behaviour by Power and Exponential Cost Functions**

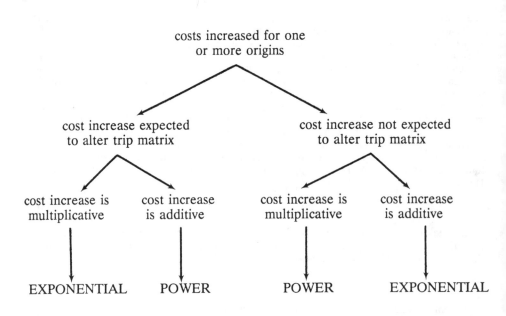

The third issue in assessing the usefulness of the two cost functions is a much simpler one to understand and concerns the behaviour of the two functions as costs tend to zero. In the case of the power function, as $c_{ij}$ tends to zero, $f(c_{ij})$ tends to infinity which is obviously problematic, whereas in the exponential function, as $c_{ij}$ tends to zero, $f(c_{ij})$ tends to one. Thus, a common finding is that the power function overestimates low cost (short distance) movements and, as a consequence, overestimates high cost (long distance) movements.

The fourth issue is an interesting finding of Choukroun (1975) concerning the aggregation of individuals in a trip matrix. Trip matrices consist of the aggregation of a large number of individual movements. Choukroun makes the point that use of the negative exponential function is consistent with the assumption that there is a constant distance-decay parameter for all trip-makers in the system. Thus, the exponential function will be appropriate when the trip-makers on which the model is calibrated are relatively homogeneous. The inverse power function, on the other hand, is more appropriate when there is a heterogenous group of trip-makers and where the distance-decay parameters for the individuals are distributed according to a gamma distribution.

In practice, the debate over the form of the cost function in spatial interaction models has evolved to a reasonably widespread consensus that the exponential function is more appropriate for analysing short distance interactions such as those that take place within an urban area. The power function, conversely, is generally held to be more appropriate for analysing longer distance interactions such as migration flows.  In the subsequent applications of spatial interaction models we will generally follow this consensus. Empirical comparisons of the behaviour of power and exponential functions can be found in Deacon et al. (1972) and in Tomlin and Tomlin (1968).

In this first chapter we have introduced the basic ideas and concepts underlying spatial interaction modelling.  We have presented a brief overview of ways in which spatial interaction models can be of use in the analysis of urban and regional systems and have discussed some of the fundamentals of operationalising such models.  Finally, a very general form of spatial interaction model was presented.  We now turn to the derivation of several specific forms of interaction models that will be applied in later chapters.

# CHAPTER 2

## A FAMILY OF SPATIAL INTERACTION MODELS

### 2.1 Introduction

Since the pioneering work of Wilson (1967; 1974) it is customary to refer to the wide range of different gravity models as belonging to a "family"; the distinguishing features of the family members is that they can be derived from a consistent optimisation problem. In the initial phases of the development of this field the optimisation problem was often that of "entropy maximisation". Later Webber (1979), March and Batty (1975a; 1975b), and Snickars and Weibull (1977) encouraged spatial interaction theorists to use a more general approach based on information minimisation. There are strong connections between the two techniques, and many researchers have drawn attention to the overlap (examples include Walsh and Webber, 1977 and O'Kelly, 1981a). The information-minimisation approach is more general and has advantages in terms of its adaptability to continuous spatial measures (see for example Bussiere and Snickars, 1970, and Angel and Hyman, 1976). For these reasons the models developed below will be presented in an information-minimising (IM) form. The following section reviews the fundamental combinatorial results which lie at the heart of the IM method. The technique will then be put to use to develop the conventional members of the family of spatial interaction models, further descriptions of which can be found in Batty (1976a), Senior (1979) and Haynes and Fotheringham (1984). These models include formulations which are useful for retail trade area analysis (see also Chapter 6) and distribution of households around fixed places of work. The formal optimisation approach also reappears in Chapter 8 which combines spatial interaction and location-allocation models.

Following this, a framework attributable to Alonso (1973; 1978) is described in which each specific member of the family of spatial interaction models can be derived by selecting particular combinations of parameters in a general model. Use is made of some recent results by Fotheringham and Dignan (1984) to show that an infinite number of spatial interaction models can be produced from the Alonso formula. Fotheringham and Dignan's extension of the Alonso framework is shown to be similar in intent to the formulation of a set of "relaxed" spatial interaction models which allow flexibility in the marginal totals of the predicted flow matrix. The formulation of relaxed interaction models is given and some results from Hallefjord and Jornsten (1985) are described.

Finally, a description is given of a slightly different framework for deriving a family of spatial interaction models developed by Tobler (1983). This family is characterised by having additive rather than multiplicative balancing factors. Throughout this chapter two types of example are given: first in each section there are references to exemplary applications of the model, and second there are short notes on sample numerical calculations. The numerical examples are necessarily abbreviated because we cannot report the full data sets, but in many instances references to complete data sources are provided.

### 2.2 Background to Information-Minimising Techniques

The "states" of a system are a set of mutually exclusive and collectively exhaustive categories into which we wish to classify a sample or a population. In the following set of notes a "macrostate" is a description which gives the numbers of each kind of item in each state of a system. A "microstate" gives the names of the items which are of each kind in each state.  For example let there be 4 individuals W, X, Y, Z.  A typical macrostate description of these individuals would say that two are in category A and two are in category B.  A microstate description would name the pairs.

The main features of the statistical mechanics as a basis for the gravity model are now outlined (for elementary reviews see Gould, 1972; Cesario, 1975a).  The two basic principles from statistical mechanics which are of use are (Georgescu-Roegen, 1971, 142):

A. The disorder of a microstate is ordinally measured by that of the associated macrostate (i.e. the macrostate to which the microstate can be aggregated).

B. The disorder of a macrostate is proportional to the number of the microstates which correspond to that macrostate.

These concepts will now be used to give an intuitive exposition of the notion of entropy and a brief justification for the formalism due to Jaynes (1957) which states that in choosing a probability distribution over the states of a system, one should do so in order to maximise entropy.  In the following discussion two cases will be considered: (1) equal categories, (2) unequal categories.  In general there will be N individuals and M categories or system states.

## 2.2.1   The Equal Categories Case

Suppose that there are 4 individuals W, X, Y and Z.  There are 5 possible ways of placing these individuals into 2 categories: (4,0), (3,1), (2,2), (1,3) and (0,4) where the ordered pairs show the numbers of individuals in category 1 and category 2.  In terms of the definitions given above these are macrostates.  The problem is to pick that macrostate which has the highest number of corresponding microstates.  The formula for the number of microstates associated with each macrostate is in general

$$R \;=\; N\,!\,/\,\textstyle\prod_i n_i\,! \;. \tag{2.1}$$

The principles stated above indicate that this is an ordinal measure of the disorder of any of the microstates as well as of the macrostate (Georgescu-Roegen, 1971, 143), and that the disorder of each of the macrostates is proportional to the number of their corresponding microstates.  The maximum entropy formalism states that the distribution which maximises R should be chosen for the positive reason (Jaynes, 1957) that it is the one which is consistent with known information and yet maximally noncommittal with respect to missing information (see also Levine and Tribus, 1979).  In the absence of any further information it is well known that the Jaynes formalism assigns equal probabilities to each of the outcomes.  This is the outcome which has maximum uncertainty, and unless other data are available to modify the distribution, we must choose this as the least biased estimate.

To give a clearer picture of the objective function in this mathematical program it is convenient to take logarithms and divide by N (this will not affect the maximum):

$$H_1 = (1/N)\log R \;=\; (1/N)(\log N! - \textstyle\sum_i \log N_i!) \tag{2.2}$$

where $N_i$ is the number of entries in the category i (see Walsh and Webber, 1977, 402). The measure represents the mean uncertainty about the occurrences of any event in a completely sampled population relative to some known facts about that population and has been deduced from a sampling without replacement process by Walsh (1976).  If all $N_i$ are large, use can be made of Stirling's approximation, $\log x! = x\log x - x$ to obtain the approximation to $H_1$:

$$H_1 = (1/N)\ (N\log N - N - \sum_i N_i \log N_i + \sum_i N_i\ )$$

$$= \log N\ - (1/N) \sum_i N_i \log N_i\ = -\sum_i (N_i/N)\log(N_i/N)\ . \tag{2.3}$$

The form of this equation is identical to that used to estimate Shannon's (1948) entropy measure,

$$H_2 = -\sum_i p_i \log p_i\ , \tag{2.4}$$

where $N_i/N$ is used to estimate $p_i$ (i = 1, ..., M).  Shannon's (1948) measure $H_2$ is one which satisfies certain reasonable requirements of a measure of uncertainty [see Webber (1977) for an outline of these requirements].

The formalism due to Jaynes (1957) states that in picking a probability distribution (in the absence of complete information) the $p_i$ values should be chosen to maximise $H_2$.  It has been found that the principle of maximum entropy is a useful tool for building models of urban phenomena. The point of departure for these urban models has been the problem of defining the probability of occurrence of discrete events. Typically the "events" are things like membership in a particular journey to work category, or membership in the trade area of a particular shopping center. To calculate these probabilities with accuracy requires a great deal of information which is not ordinarily available. Thus there is a need for a method to allow an estimate of the group probabilities to be found. That such a method is needed is unquestionable since as Jaynes puts it "the amount of information available in practical situations is so minute that it alone would never suffice for making reliable predictions" (Jaynes, 1957, 625).  Therefore Jaynes states that in choosing a probability distribution one should maximise entropy subject to constraints embodying whatever limited information is known. This formalism has been accepted as a powerful tool (for examples see Tribus, 1969; Dowson and Wragg, 1973) and has found many uses in human geography; Webber (1975, 101) lists several possible geographical applications.

## 2.2.2  The Unequal Categories Case

It is also possible to work with categories which are not a priori equally likely.  As an example begin by placing 4 individuals W, X, Y and Z into 2 unequal categories:  A (which has two cells) and B (which has one cell).  Thus the following ordered pairs show the possible allocations of the objects into two categories: (4,0), (3,1), (2,2), (1,3), (0,4).

The number of microstates giving rise to the macrostates is known from $H_1$ above. Then for each of these microstates there are $2^{N_A}$ ways of distributing the $N_A$ individuals in category A over its two cells, and obviously just one way of placing the individuals in category B into its single cell.

In general for N items and for M categories, the total number of microstates that correspond to a given macrostate is

$$V = [N!/(N_1!\ N_2!\ ...\ N_M!)].[S_1^{\,N_1} ...\ S_M^{\,N_M}] \tag{2.5}$$

where $S_i$ is the number of cells in the $i^{th}$ category, and $N_i$ is the number of entries in the $i^{th}$ category. (Notation and methods follow Snickars and Weibull, 1977.) The numbers of microstates which arise in the equal categories case and the number of fine grain microstates which arise in the unequal categories case are summarised in Table 2.1 for a small numerical example.

**Table 2.1  Microstate (Equal Categories) and Fine Grain Microstates (Unequal Categories)**

| Macrostates | 4,0 | 3,1 | 2,2 | 1,3 | 0,4 |
|---|---|---|---|---|---|
| number of microstates | 1 | 4 | 6 | 4 | 1 |
| number of fine grain microstates | 16 | 32 | 24 | 8 | 1 |

Examining this table it is clear that the macrostate distribution with greatest number of microstates is (2,2) while the macrostate distribution with the largest number of fine grain microstates is (3,1). Therefore it is important that any information on category size be incorporated into an entropy-maximising model. (This point is also raised by Batty (1976b, 3).)

To show what is being maximised in the expression of V, take logs, divide by N and subtract log M.

$$H_3 = (1/N)\log V - \log M =$$

$$= (1/N)(\log N! + \sum_i N_i \log S_i - \sum_i N_i!) - \log M. \tag{2.6}$$

Using Stirling's approximation

$$H_3 = (1/N)(N\log N + \sum_i N_i \log S_i - \sum_i N_i \log N_i) - \log M =$$

$$= \log(N/M) - (1/N)\sum_i N_i \log(N_i/S_i) . \tag{2.7}$$

Rearranging and taking the negative gives

$$- H_3 = \log M + \sum_i (N_i/N)\log(N_i/N.S_i) \tag{2.8}$$

Equation (2.8) has been called negentropy by Walsh and Webber (1977, 414), and inverse spatial entropy by Batty (1976b, 4). Finally combining terms gives the following expression for Kullback information gain, (KIG):

$$- H_3 = KIG = \sum_i (N_i/N)\log(N_i/N)/(S_i/M). \tag{2.9}$$

The form of this equation is identical to that used to calculate

$$KIG = \sum_i p_i \log(p_i/q_i) , \tag{2.10}$$

where $N_i/N$ is used to estimate $p_i$ ($i = 1,..., M$) and $S_i/M$ is used to estimate $q_i$ ($i = 1,..., M$), where $q_i$ is a prior probability of membership in category i. If the priors are unequal, Hobson (1969) has shown that under certain reasonable conditions KIG is a unique (up to a multiplicative constant) measure of the information gained when a distribution $p_i$ ($i = 1,...,$ M) replaces a prior $q_i$ ($i = 1,...,$ M). This measure is discussed in more detail by Hobson (1969) and Hobson and Cheng (1973).

KIG can be accepted as a measure of information gain for cases where the categories (henceforth used interchangeable with 'priors') are unequal. An extension of Jaynes' formalism, called the <u>minimum information principle</u>, has been suggested by Evans (1969), Hobson and Cheng (1973) and in a geographical context by Snickars and Weibull (1977). This principle states that the distribution which minimises information gain should be chosen in order to obtain an estimate of the probabilities $p_i$ from limited information, and in the presence of prior information encoded in $q_i$. From an information theory point of view Kullback's uncertainty should be maximised subject to constraints. Kullback uncertainty is defined as the difference between the largest information gain possible and the information gain actually achieved, that is:

$$KU = \sum_i p_i^{max}\log (p_i^{max}/q_i) - \sum_i p_i\log (p_i/q_i), \qquad (2.11)$$

where $p_i^{max}$ is the distribution representing maximum information consistent with the constraints (i.e. a zero-one distribution). Since the first term on the right does not depend on $p_i$ it produces an irrelevant constant. The modified formalism is therefore to maximise KIG (or $H_3$). If the prior distribution is uniform then the term being maximised is again Shannon's entropy ($H_2$) (Hobson and Cheng, 1973, 305; Tribus and Rossi, 1973, 335; Snickars and Weibull, 1977, 146). There has been some discussion of the modified formalism in a geographical context (see Webber 1979 for a thorough review). March and Batty (1975b) demonstrate that there is a class of minimally-prejudiced models of which the Kullback and Shannon based formalisms are special cases. Applications of the modified formalism include March and Batty (1975a), Batty and March (1976) and Snickars and Weibull (1977).

The notation defined in the previous paragraphs needs one final refinement in order to make it applicable to the spatial interaction context. Instead of indexing the categories by just a single subscript (e.g. $p_i$) it is better to use two subscripts for the cells in a matrix. Thus the entropy and information gain statistics are denoted with $p_{ij}$ values: the i subscript denotes the rows of the table while the j subscript denotes the columns. In terms of the earlier notation: if there are n origins and m destinations then there are nm boxes in the macro state space (see Snickars and Weibull, 1977, 152).

Cesario (1975a) gives a useful numerical example which will serve to illustrate the entropy-maximising formalism and the modified formalism based on prior information:

Table 2.2  **Example Interaction System**

|                    |     | Destinations | |                |
|--------------------|-----|--------------|----------|-----------------|
|                    |     | 1            | 2        | Sum of outflows |
| Origins            | 1   | $T_{11}$     | $T_{12}$ | 3               |
|                    | 2   | $T_{21}$     | $T_{22}$ | 3               |
| Sum of inflows     |     | 4            | 2        |                 |

The following notation is used: suppose that n zones interact and that the measurement of these interaction levels is recorded in a matrix **T** with elements $T_{ij}$. (The number of origins and destinations is the same in this example - this does need to be true in general.) The only "information" which we use to constrain the interaction matrix **T** is the sum of inflows to the destinations and the sum of outflows from each origin. Assume further that only integer combinations are considered.  Consider the spatial interaction matrix in Table 2.2. The problem is to find the distribution $T_{ij}$ which is maximally noncommittal with respect to missing information and at the same time consistent with the constraints on the inflows and outflows.  The possible macrostates which are consistent with the information are:

(a)    T =    3   0          (b)    T =    2   1          (c)    T =    1   2
              1   2                        2   1                        3   0

The total number of ways we can select a particular distribution $\{T_{ij}\}$ from T using no prior information about the probability of the individual linkages is:  $R = T!/(T_{11}! \; T_{12}! \; T_{21}! \; T_{22}!)$ . If R is evaluated for each of the above distributions, it is found that in (a) $R = 60$, in (b) $R = 180$, and in (c) $R = 60$.   The most likely distribution is the one with the most microstates associated with it - hence (b) is chosen.

To extend this example to the unequal categories case suppose that the number of cells in each category is  $S_{11} = 2$, $S_{12} = 1$, $S_{21} = 1$, and $S_{22} = 2$.  Then the number of fine grain microstates consistent with each macrostate is given by:

$$V = \frac{T! \; (\prod_i \prod_j S_{ij}^{\;T_{ij}})}{\prod_i \prod_j T_{ij}!} . \tag{2.12}$$

The results for the above values of $S_{ij}$ are (a) 1920, (b) 1440 and (c) 120.  Thus (a) is now the macrostate with the greatest number of associated microstates.  Snickars and Weibull (1977) suggest that prior information in the trip distribution context could consist of an interaction matrix from another period.  One possible interpretation of the prior information in this context is that the number of routes between the zones is an appropriate piece of information that could be incorporated.  In the above case this would mean that there are two routes from zones 1 to 1, and 2 to 2; and, only one route from 1 to 2, and 2 to 1. It is obviously important to use prior information whenever it is available because the solution interaction matrix is sensitive to the choice of these data. Note that the prior data may literally be an interaction table from a previous period. This is the approach used by Snickars and Weibull (1977). On the other hand more theoretical models such as that used by Webber and O'Kelly (1981) simply postulated prior probabilities based on assumptions about the likely occurrence of different events. The modeller has to make a decision whether to use prior data as a weight (as in the KIG measure) or whether to use these data as actual constraints on the interaction system. Be aware that the former use does not necessitate a consistency between the prior data and the current period's constraints.

In conclusion, the two important measures (2.3) and (2.9) from this section are written out in the trip distribution context:

$$H_2 = - \sum_i \sum_j (T_{ij}/T) \log (T_{ij}/T) , \tag{2.13}$$

$$KIG = \sum_i \sum_j (T_{ij}/T) \log (T_{ij}/T)/(S_{ij}/M) . \tag{2.14}$$

where $H_2$ is the entropy measure of the distribution $T_{ij}/T$, and KIG is the Kullback information gain for the trip distribution matrix $T_{ij}/T$ with the prior probability matrix $S_{ij}/M$. (Obviously the sum over all the $T_{ij}$ values is T, and the sum over all the $S_{ij}$ values is M.) A strict adherence to the convention that $H_2$ and KIG statistics are reported numerically as functions of the flows (and priors) <u>expressed as proportions</u> is maintained in this book. Note however that in the usage of these measures in maximum-entropy or minimum-information models, the scaling is irrelevant to the solution. That is, it is possible to work with unscaled values of $T_{ij}$ and $S_{ij}$ in (2.13) and (2.14) and obtain the same equations for the solution as in the scaled version, since the objective differs only by a constant from the scaled version. It is important to clarify the units of the $T$ matrix when discussing the constraints: when the elements of $T$ are scaled to sum to one, then the row and column constraints are expressed as proportions, and the trip length constraint takes the form of a system average; however, when the values are unscaled the row and column totals are in units of flow, and the trip length constraint is based on total interaction volumes. The program SIMODEL (Williams and Fotheringham, 1984, and Chapter 3 below) follows the convention that entropy and information statistics are computed from flows expressed as proportions, while the flow units themselves can be scaled or unscaled according to the modeller's choice.

## 2.3 Information Concepts and the Family of Interaction Models

The techniques that are described in the previous section are now used to derive a series of different spatial interaction models - with and without the use of prior information. Four types of model are considered in the subsections 2.3.1 - 2.3.4: (1) Unconstrained; (2) Production-constrained; (3) Attraction-constrained; and (4) Doubly constrained. Within each of the subsections consideration is given to simple entropy-maximising formulations (based on objective (2.13)) and information-minimising versions using explicit non-uniform priors in objective (2.14).

The various types of models which can be derived using either formalism are distinguished by the constraints which are embodied in them. Recalling the discussion in Chapter 1, four cases can be noted, (O=origin, D=destination):

(1)  Neither the set of row totals $O_i$ nor the set of column totals $D_j$ is known,

(2)  The set of row totals $O_i$ is known,

(3)  The set of column totals $D_j$ is known,

(4)  Both sets of marginal totals are known.

Therefore in the following models zero, one or two of the following constraints may operate:

$$\sum_j T_{ij} = O_i , \tag{2.15}$$

$$\sum_i T_{ij} = D_j . \tag{2.16}$$

In addition a trip length constraint is imposed on the interactions. This constraint requires that the model reproduce some observed total trip length, C. (This constraint is sometimes reworked as an average trip length constraint. It is interesting that higher moments are not usually constrained in interaction models, although the technique for this adaptation is readily available: Dowson and Wragg, 1973; Tribus 1969.) The trip length constraint has the form:

$$\sum_i\sum_j T_{ij} d_{ij} = C , \tag{2.17}$$

where $d_{ij}$ is the distance between i and j, and C is the total system trip length (or trip time). Wilson (1974) has produced a family of spatial interaction models using these constraints. His results are now generalised using the Kullback-based formalism.

2.3.1  Unconstrained Models

The result of the maximisation of (2.13) subject to (2.17) produces

$$T_{ij} = \exp(-\beta d_{ij}) \, , \tag{2.18}$$

where $\beta$ is found so as to satisfy (2.17). Cordey-Hayes and Wilson (1971) introduce the origin and destination attraction factors through the existence of savings in costs associated with certain places.    An alternative method for introducing these factors in the unconstrained case is to minimise a function, based on equation (2.14), of the form:

$$\Sigma_i \Sigma_j \, T_{ij} \, \log \, (T_{ij}/v_i^\mu w_j^\alpha) \, , \tag{2.19}$$

where $v_i$ and $w_j$ incorporate some prior knowledge of the attraction of origins and destinations. In (2.19) $\mu$ is a parameter reflecting the relationship between $T_{ij}$ and $v_i$ while $\alpha$ reflects the relationship between $T_{ij}$ and $w_j$. The result of this optimisation is an expression

$$T_{ij} = v_i^\mu w_j^\alpha \, \exp \, (-\beta d_{ij}) \, , \tag{2.20}$$

which is the form of the traditional "unconstrained" gravity model. (Of course there is a constraint implied by the average trip length restriction, but the model is unconstrained in terms of the production of trips from origins and the attraction of trips to destinations.)

Many early gravity model formulations are reported in the format of unconstrained models (see Haynes and Fotheringham, 1984). Following the rationale in Chapter 1 we discount the usefulness of these models, because of the relatively poor quality of the predictions derived from them, and move directly to the development of production-, attraction-, and doubly constrained models.

2.3.2  Production-Constrained Models

When independent estimates of the numbers of flows originating in each zone are known this information must be built into the estimates of $T_{ij}$. This is achieved by maximising (2.13) subject to (2.15) and (2.17) then the resulting expression is

$$T_{ij} = A_i O_i \, \exp \, (-\beta d_{ij}) \, , \tag{2.21}$$

where

$$A_i = (\Sigma_i \, \exp(-\beta d_{ij}))^{-1} \, . \tag{2.22}$$

The $A_i$ variable is called the partition function in information theory, (see Tribus, 1969, 124). The value of $A_i$ serves to ensure that the model reproduces the volume of flow originating at zone i. It is often referred to as a balancing factor in the spatial interaction literature.

It is possible to introduce prior information on the attraction of various destination zones. The interpretation is that although the model accurately reflects the outflow from each origin, there is no representation of the attraction of each destination. The problem is formulated so as to minimise

$$\Sigma_i \Sigma_j \, T_{ij} \, \log \, (T_{ij}/v_i^\mu w_j^\alpha) \, , \tag{2.23}$$

where $v_i$ is arbitrary (uniform) and $w_j$ is non-uniform and reflects the attraction of a destination; subject to (2.15) and (2.17). The result is

$$T_{ij} = A_i w_j^\alpha O_i \, \exp(-\beta d_{ij}) \qquad\qquad (2.24)$$

where now

$$A_i = (\sum_j w_j^\alpha \exp(-\beta d_{ij}))^{-1} \; . \qquad\qquad (2.25)$$

In terms of the application of production-constrained models there has been a major area of activity, namely, the allocation of retail flows from residential zones to retail establishments and this type of application is discussed in detail in Chapter 6.

Production-Constrained Model - Numerical Example

A 17 x 17 interaction system representing flows from origin zones to retail attractions in Hamilton (Ontario) provides a test case for the production-constrained model given in equation (2.24). The model is fitted using grocery shopping trips as the observed data. The results are discussed in the following paragraph and in Table 2.3, and later in Chapter 7, Table 7.7. The values in the observed interaction matrix are based on the aggregated interaction between each origin zone and each destination; in other words, the flows reflect all trips which include any interaction between the origin base and the destination. The significance of this is that many diverse trips are being bundled together for the purposes of this simple analysis. As will be discussed later in the book, it is usually necessary to distinguish between simple single purpose trips and those which involve interactions between many destinations and purposes. Notwithstanding these qualifications, it appears that grocery shopping trips can be represented quite well with a production-constrained interaction model. Many aspects of these data accord with intuition about shopping trips: the average trip length for the grocery sample is 5.1492 minutes yielding a $\beta$ parameter of 0.4456. (The longer and more dispersed nature of non-grocery spatial interaction has been documented in previous reviews of urban travel (see Chapter 7)). The exponent on the attraction term is 1.23076. The destination inflows are quite accurately matched by the models reflecting the commonly observed ability of this model to replicate aggregate flows reasonably well. However, when the individual origin-destination pairs are examined, several major errors are evident. A major component of travel which is unaccounted for in this analysis, namely multipurpose trip-making, is discussed in Chapter 7.

2.3.3   Attraction-Constrained Models

This case is the mirror image of the production-constrained model. $D_j$ is now known independently, and the problem therefore is to maximise (2.13) subject to (2.16) and (2.17). The result of this is

$$T_{ij} = B_j D_j \, \exp(-\beta d_{ij}) \; , \qquad\qquad (2.26)$$

where

$$B_j = (\sum_i \exp(-\beta d_{ij}))^{-1} \; . \qquad\qquad (2.27)$$

Wilson adds in factors to represent the various trip origin zones (based on residential land use) by subtracting an origin-specific constant from the cost of each flow. This effect can also be achieved by minimising

$$\sum_i \sum_j T_{ij} \log (T_{ij}/v_i^\mu w_j^\alpha) \; , \qquad\qquad (2.28)$$

**Table 2.3 Production-Constrained Model, Predicted and Observed Trips Ends at the 17 destinations; Grocery trips**

| J | OBSERVED INFLOWS | PREDICTED INFLOWS |
|---|---|---|
| 1 | 38.0 | 26.4 |
| 2 | 40.0 | 17.7 |
| 3 | 269.0 | 240.1 |
| 4 | 398.0 | 429.3 |
| 5 | 205.0 | 150.3 |
| 6 | 429.0 | 433.8 |
| 7 | 156.0 | 162.4 |
| 8 | 161.0 | 160.4 |
| 9 | 353.0 | 428.6 |
| 10 | 413.0 | 366.0 |
| 11 | 375.0 | 387.6 |
| 12 | 30.0 | 28.7 |
| 13 | 92.0 | 124.9 |
| 14 | 62.0 | 92.6 |
| 15 | 223.0 | 195.1 |
| 16 | 334.0 | 291.6 |
| 17 | 191.0 | 233.7 |

Source: Authors' calculations using SIMODEL Option 2, a two-parameter production-constrained interaction model.

where $v_i$ is non-uniform and $w_j$ is arbitrary (uniform), subject to (2.16) and (2.17) and the result is

$$T_{ij} = v_i^\mu B_j D_j \exp(-\beta d_{ij}) , \qquad (2.29)$$

and now

$$B_j = (\sum_i v_i^\mu \exp(-\beta d_{ij}))^{-1} . \qquad (2.30)$$

The major application of attraction-constrained models has been to location of residential land use. The significance of an attraction-constrained model is that the modeller can use information on (say) employment as an indicator of work trip attraction together with trip length data to estimate a trip matrix. Then, summing the elements in each row i of this matrix will give an estimate of trip productions from origin zone i. These can then be interpreted as the assignment of the workers from their place of work to zones of residence. As was shown above the attraction-constrained model is of the form

$$T_{ij} = v_i^\mu B_j D_j f(d_{ij}) , \qquad (2.31)$$

where $B_j$ is calculated to ensure that the trips ending in zone j match the independent information. Since there is no restriction on the number of trips originating at i the model can be said to estimate the amount of activity located in i. Thus the simplest modification of the Lowry model (Wilson, 1969) which allocates workers to zones of residence is

$$P_i = v_i^\mu \sum_j B_j D_j \, f(d_{ij}), \tag{2.32}$$

where $P_i$ is an estimate of the total allocation workers to zone i as a place of residence. This removes the inconsistency mentioned in early critiques of the Lowry model, (Broadbent, 1970, 469-70). Further developments in residential location modelling are reported in Wilson (1975) and in Wilson, Coehlo, Macgill and Williams (1981, Chapter 6).

A final example of this type of model is provided in Mayhew and Leonardi (1982) in the area of hospital planning. Their idea is to use a planned set of hospital resource allocations to describe the available destinations for health care. The distribution of the expected users of these fixed facilities is derived from an attraction-constrained interaction model: since the model allows assignment to an unsatisfied demand category, there is no strict constraint on the numbers of users emanating from each district. Therefore by comparing the potential demand in each district to the amount estimated to be serviced there, the model can provide a measure of the equity or efficiency of the health care system. The entire model is embedded in a prescriptive framework by allowing the hospital resources to be a function of system performance.

## Attraction-Constrained Model - Numerical Example

As an application of the attraction-constrained model, a practical problem which arises in evaluating places of residence of a set of factory workers is tackled. The problem derives from the following circumstances: several inner city factories in Canton, Ohio are located in an area which is designated as an Enterprise Zone (EZ) under the State of Ohio's EZ plan. While the terms of the EZ legislation entitles the factory to some tax relief, the ultimate impact is felt in the area in which the factory workers reside. A sample of the factories was surveyed in order to determine the residential location of their employees. The reason that it is of interest to determine the likely places of origin of these workers, is that the enterprise zone plan attempts to promote further employment expansion in the EZ: if it is assumed that any new employment created adds workers with similar journey to work characteristics as the current workforce, then it is possible to predict the impact of the enterprise zone by place of residence (something which cannot be done without such a distributional technique).

The model to illustrate the attraction-constrained interaction design is as follows:   (1) Survey data from twenty factories show the number of employees and an estimate of their length of trip to work. These observations provide data for the trip end constraints and the destination specific trip length parameter. (2) Readily available census data about journey to work length can be used also - since the relative distance between the factories and the census tract origins is known, then only those residents of the census tract who have a trip length compatible with this distance are used as the weights for the origin. For instance a tract might have 2000 employed persons over the age of 16, distributed over 8 different journey to work categories. Only those in the category which is consistent with the spatial separation between that census tract and the inner city destination need to be used. Of course, these employment levels enter only as weights (not constraints). The necessary data have been collected for Canton Ohio and Table 2.4 shows the salient data for some of the 84 tracts. The table shows the tract identifier, the average distance from each tract to the central city factory districts (themselves made up of several tracts), and the estimated number of persons who are in journey categories consistent with the spatial separation between the tract and the central city work place (assuming an average speed of 30 mph). Finally, the model's prediction of the number of factory workers who are estimated to come from each tract is shown.

## Figure 2.1 Canton Census Tracts: Predicted Origins for Central City Employees

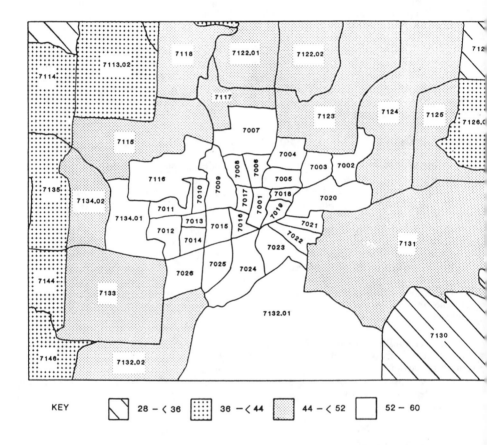

Predicted number of employees in tract as a percentage of the number of
employees in the tract who travel appropriate average distance to work

**Table 2.4  Tract ID, and selected model results**

| ID | TRACT | D1 | D2 | D3 | ORIGIN | PRED |
|---|---|---|---|---|---|---|
| 1 | 7001.00 | 1.11 | 1.43 | 1.68 | 123.5 | 73.14 |
| 2 | 7002.00 | 1.02 | 1.53 | 1.70 | 106.0 | 62.54 |
| 3 | 7003.00 | 1.25 | 3.12 | 2.65 | 104.5 | 56.10 |
| 4 | 7004.00 | 1.71 | 3.00 | 3.02 | 417.0 | 221.11 |
| 5 | 7005.00 | 1.00 | 2.30 | 2.20 | 144.5 | 81.36 |
| 6 | 7006.00 | 1.62 | 2.10 | 2.53 | 174.0 | 97.07 |
| 7 | 7007.00 | 2.58 | 2.63 | 3.37 | 809.5 | 427.52 |
| 8 | 7008.00 | 1.86 | 1.89 | 2.54 | 191.5 | 107.45 |
| 9 | 7009.00 | 2.32 | 2.16 | 2.96 | 109.0 | 59.50 |
| . | . | . | . | . | . | . |
| . | . | . | . | . | . | . |
| . | . | . | . | . | . | . |
| 71 | 7139.00 | 9.56 | 7.53 | 8.96 | 315.0 | 113.97 |
| 72 | 7140.00 | 10.59 | 8.53 | 9.95 | 683.5 | 231.53 |
| 73 | 7141.00 | 9.50 | 7.43 | 8.85 | 674.0 | 245.49 |
| 74 | 7142.00 | 8.77 | 6.55 | 7.88 | 356.5 | 137.98 |
| 75 | 7143.01 | 8.41 | 6.36 | 7.78 | 414.0 | 162.10 |
| 76 | 7143.02 | 7.16 | 5.14 | 6.57 | 585.0 | 249.20 |
| 77 | 7144.00 | 7.48 | 5.24 | 6.53 | 346.5 | 146.97 |
| 78 | 7145.00 | 9.47 | 7.21 | 8.47 | 31.5 | 11.67 |
| 79 | 7146.00 | 9.59 | 7.33 | 8.32 | 851.0 | 314.76 |
| 80 | 7147.01 | 11.38 | 9.30 | 10.71 | 975.5 | 314.40 |
| 81 | 7147.02 | 13.80 | 11.60 | 12.95 | 506.5 | 141.26 |
| 82 | 7148.00 | 14.37 | 12.14 | 13.01 | 1756.0 | 476.84 |
| 83 | 7149.01 | 8.28 | 7.57 | 6.94 | 760.0 | 289.90 |
| 84 | 7149.02 | 9.95 | 10.32 | 9.13 | 402.0 | 129.09 |

ID:      Identifier for sequence number of tracts
TRACT:   Tract number

D1:      Mileage to factory district 1 (comprises tracts
         7018, 7019, 7020 and 7021)
D2:      Mileage to factory district 2 (comprises tracts
         7001, 7014-7016, 7025 and 7026)
D3:      Mileage to factory district 3 (comprises tracts
         7022, 7023 and 7024)

ORIGIN:  Number workers who travel this average
         distance to work

PRED:    Predicted number of factory employees in tract

Source: Master Area Reference File for the State of Ohio, 1980 Census of Population, and authors' calculations.

## 2.3.4  Doubly Constrained Model

This problem involves maximising (2.13) subject to (2.15), (2.16) and (2.17). The result of the constrained maximisation is that

$$T_{ij} = A_i B_j\, O_i D_j\, \exp(-\beta d_{ij}) \,, \qquad (2.33)$$

and

$$A_i = (\Sigma_j\, B_j D_j\, \exp(-\beta d_{ij}))^{-1} \,, \qquad (2.34)$$

$$B_j = (\Sigma_i\, A_i O_i\, \exp(-\beta d_{ij}))^{-1} \,. \qquad (2.35)$$

where $A_i$ and $B_j$ are interrelated balancing factors which ensure that the constraints on origins and destinations are met.  An obvious extension of this result is to incorporate prior information on the distribution $T_{ij}$.  That is minimise (2.14) where $S_{ij}$ is some existing flow data subject to (2.15), (2.16) and (2.17).  The result is an expression of the form

$$T_{ij} = S_{ij}\, A_i B_j\, O_i D_j\, \exp(-\beta d_{ij}) \,, \qquad (2.36)$$

and now

$$A_i = (\Sigma_j\, S_{ij}\, B_j D_j\, \exp(-\beta d_{ij}))^{-1} \,, \qquad (2.37)$$

$$B_j = (\Sigma_i\, S_{ij}\, A_i O_i\, \exp(-\beta d_{ij}))^{-1} \,. \qquad (2.38)$$

As a special case let $S_{ij} = v_i{}^{\mu} w_j{}^{\alpha}$ for all i and j.  Then the problem is to minimise

$$\Sigma_i \Sigma_j\, T_{ij}\, \log\, (T_{ij}/v_i{}^{\mu} w_j{}^{\alpha}) \,, \qquad (2.39)$$

subject to (2.15), (2.16) and (2.17). The result is an expression of the form

$$T_{ij} = A_i B_j\, O_i D_j\, \exp(-\beta d_{ij}) \,, \qquad (2.40)$$

which is identical to (2.33); the appropriate balancing factors are

$$A_i = (\Sigma_j\, w_j{}^{\alpha}\, B_j D_j\, \exp(-\beta d_{ij}))^{-1} \,, \qquad (2.41)$$

$$B_j = (\Sigma_i\, v_i{}^{\mu}\, A_i O_i\, \exp(-\beta d_{ij}))^{-1} \,. \qquad (2.42)$$

This model is a direct generalisation of (2.33) above. Applications of it have been provided by Snickars and Weibull (1977) and the appropriate solution method has been outlined in Karlqvist and Marksjo (1971).

The doubly constrained model has found a wide applicability in trip distribution problems. In these cases it is usually known that certain numbers of trips originate and end in each zone. Entropy-maximising estimates of the detailed interzonal flows are easily obtained. Extensions have included combination of the modal split and trip distribution stages of the transportation planning process, by simply adding a subscript for each mode and by using information on average trip length by mode. Much more elaborate and theoretically challenging enhancements have blended the trip assignment and trip distribution phases of the transportation planning process (see Evans, 1976 and Jornsten, 1980). A brief statement of the Jornsten model is described here, in the same notation as the previous section.

The model for combined trip distribution and assignment aims to choose a trip distribution matrix to maximise entropy, while combining the allocations between zones over a number of alternative routes. That is, choose $T_{ij}$ so as to maximise:

$$- \sum_i \sum_j T_{ij} \log T_{ij} \qquad (2.43)$$

subject to

$$\sum_j T_{ij} = O_i, \qquad (2.44)$$

$$\sum_i T_{ij} = D_j, \qquad (2.45)$$

$$\sum_k X_{ijk} = T_{ij}, \qquad (2.46)$$

$$\sum_i \sum_j \sum_k A_{ijkm} X_{ijk} \leq b_m \qquad (2.47)$$

$$\sum_m C_m \left( \sum_i \sum_j \sum_k A_{ijkm} X_{ijk} \right) \leq C^o \qquad (2.48)$$

$$X_{ijk} \geq 0, \; T_{ij} \geq 0, \; T = \sum_i \sum_j T_{ij}, \qquad (2.49)$$

where the terms are as follows:

$O_i$ = the number of journeys from origin i,

$D_j$ = the number of journeys to destination j,

$T_{ij}$ = the number of journeys from i to j,

$X_{ijk}$ = the flow on path k from i to j,

$A_{ijkm}$ = 1 if link m is part of path k between i and j,
       0 otherwise.

$C_m$ = the cost of a unit of flow on link m,

$b_m$ = the capacity of link m,

$C^o$ = the total travel cost requirement.

The idea of the model is to assign trips from origins to destinations in accordance with a conventional doubly constrained interaction model. If these flows obey the additional flow and capacity constraints, then the entire model is solved. Of course it is likely that these extra conditions are not satisfied, and then a more elaborate solution strategy is needed. It is well known that a simple iterative readjustment of the costs for the flows will not necessarily converge (see the statement and reference in Jornsten, 1980, page 264) . Related work is the early congestion model by Scott (1971) and the statement in Webber (1979, page 119) that there can be no interdependence between the flow variables and the interaction costs in the conventional entropy-maximising model. Jornsten (1980) proposes and implements a convergent procedure for the model where the appropriate travel times are built up in a series of successively more constrained problems; this idea uses the Benders decomposition method. Numerical examples are presented in Jornsten's paper.

Doubly Constrained Model - Numerical Example

As an example of the trip distribution application of the doubly constrained spatial interaction model, the data reported by Walsh (1980) for a study area in Ireland are reanalyzed. There are several different models studied in Walsh's paper but here only the 8 x 8 journey to work data for County Limerick are considered. (This example is small enough to be checked by hand calculation or using a small computer system.) The scale of the study region can be seen in Figure 2.2 which reports the interzonal travel distances and shows the zonal boundaries. The program SIMODEL (Williams and Fotheringham, 1984) which is discussed in Chapter 3 is used to calibrate the parameters of the doubly constrained model with an exponential distance function. The observed mean trip length of 7.5836 is obtained in SIMODEL by summing the observed values of the interactions, weighted by the distances. The maximum likelihood estimate of the distance parameter is 0.1909930 (which enters the model with a negative sign) obtained after 5 iterations of the calibration routine. A total of 4190 interactions were observed, and these began and ended in the origins and destinations as shown in the data in Tables 2.5 and 2.6 The predicted interactions did not exactly match the observations, and in fact an average percentage deviation of 17.4% was found. These results are not directly comparable to Walsh's due to an apparent difference in tolerance levels for convergence. Specifically, Walsh constrains the mean interaction distance to be 7.3, while our computations from the printed data in his paper suggest that the mean is 7.5836. This highlights the critical role played by the average distance constraint and the fact that the resultant model is sensitive to the numerical level of the mean value. A second difference in our results arises from the numerical accuracy of the iterative scheme for the balancing factors. Based on equations (2.34) and (2.35) $A_i$ and $B_j$ are iteratively readjusted so as to enforce simultaneous agreement with the constraints (2.15) and (2.16). The program SIMODEL seems to enforce a higher degree of accuracy than Walsh in this procedure because our predicted row and column totals match the observed values more closely.

Table 2.5   **Results of Calibration of Doubly Constrained Model**

| I | $A_i$ BALANCING FACTOR | $O_i$ OBSERVED OUTFLOW | $B_j$ BALANCING FACTOR | $D_j$ OBSERVED INFLOW |
|---|---|---|---|---|
| 1 | 0.195911E-02 | 457.0 | 0.145756E+01 | 491.0 |
| 2 | 0.885199E-02 | 92.0 | 0.185958E+01 | 70.0 |
| 3 | 0.207632E-02 | 893.0 | 0.995523E+00 | 826.0 |
| 4 | 0.135907E-02 | 208.0 | 0.114869E+01 | 243.0 |
| 5 | 0.119707E-02 | 371.0 | 0.106440E+01 | 878.0 |
| 6 | 0.210947E-02 | 1814.0 | 0.813483E+00 | 1299.0 |
| 7 | 0.270781E-02 | 160.0 | 0.259372E+01 | 232.0 |
| 8 | 0.287481E-02 | 195.0 | 0.195575E+01 | 151.0 |

Source: Authors' calculations using SIMODEL from data in Walsh (1980)

Figure 2.2  **Limerick Journey to Work Zones (Walsh, 1980, Table 2)**

```
1     4.13 28.30 16.81  6.25 17.50 19.50 22.87 31.44
2    28.30  2.99 17.49 22.07 33.30 43.60 33.50 42.10
3    16.80 17.40  4.04 10.56 21.81 32.11 20.19 28.75
4     6.25 22.07  8.50  1.72 11.25 21.55 16.62 25.19
5    17.50 33.20 21.81 11.25  2.61 10.30 13.60 22.20
6    19.50 43.20 32.11 21.55 10.30  6.39 23.90 27.25
7    22.80 33.50 20.19 16.62 13.68 23.98  5.93  8.56
8    31.40 42.12 28.75 25.19 22.24 27.25  8.56  1.94
```

Travel distances in miles

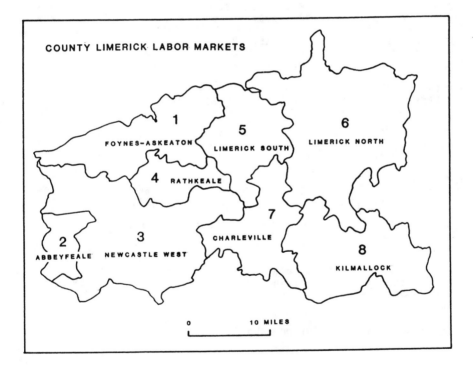

Table 2.6  **Predicted and Observed Interactions for 8 x 8 Limerick Data**

DESTINATIONS

| | | 1 | 2 | 3 | 4 | 5 | 6 | 7 | 8 | Origin outflow |
|---|---|---|---|---|---|---|---|---|---|---|
| 1 | W pred | 290. | 1. | 31. | 76. | 31. | 21. | 7. | 1. | 458 |
| | S pred | 291. | 1. | 30. | 76. | 30. | 23. | 7. | 1. | 457 |
| | obs | 351. | 1. | 24. | 25. | 39. | 12. | 5. | 0. | 457 |
| 2 | W pred | 3. | 59. | 24. | 3. | 1. | 0. | 1. | 0. | 91 |
| | S pred | 3. | 60. | 24. | 3. | 1. | 0. | 1. | 0. | 92 |
| | obs | 0. | 49. | 35. | 7. | 1. | 0. | 0. | 0. | 92 |
| 3 | W pred | 55. | 9. | 701. | 69. | 28. | 4. | 24. | 2. | 892 |
| | S pred | 54. | 9. | 705. | 69. | 27. | 4. | 24. | 2. | 893 |
| | obs | 35. | 19. | 697. | 67. | 6. | 13. | 56. | 0. | 893 |
| 4 | W pred | 61. | 1. | 46. | 56. | 32. | 4. | 7. | 1. | 208 |
| | S pred | 61. | 1. | 46. | 57. | 31. | 5. | 7. | 1. | 208 |
| | obs | 27. | 0. | 30. | 128. | 7. | 10. | 6. | 0. | 208 |
| 5 | W pred | 12. | 0. | 6. | 15. | 257. | 59. | 20. | 2. | 371 |
| | S pred | 11. | 0. | 6. | 14. | 252. | 66. | 20. | 2. | 371 |
| | obs | 23. | 0. | 19. | 8. | 253. | 59. | 9. | 0. | 371 |
| 6 | W pred | 66. | 0. | 7. | 17. | 490. | 1203. | 24. | 6. | 1813 |
| | S pred | 66. | 0. | 7. | 17. | 500. | 1193. | 24. | 6. | 1814 |
| | obs | 45. | 0. | 9. | 3. | 560. | 1167. | 18. | 12. | 1814 |
| 7 | W pred | 4. | 0. | 8. | 5. | 31. | 4. | 83. | 25. | 160 |
| | S pred | 4. | 0. | 8. | 5. | 30. | 5. | 84. | 25. | 160 |
| | obs | 9. | 1. | 10. | 5. | 10. | 14. | 95. | 16. | 160 |
| 8 | W pred | 1. | 0. | 2. | 1. | 8. | 3. | 66. | 114. | 195 |
| | S pred | 1. | 0. | 2. | 1. | 7. | 3. | 66. | 114. | 195 |
| | obs | 1. | 0. | 2. | 0. | 2. | 24. | 43. | 123. | 195 |
| $D_j$ | W pred | 492. | 70. | 825. | 242. | 878. | 1298. | 232. | 151. | 4188 |
| $D_j$ | S pred | 491. | 70. | 826. | 243. | 878. | 1299. | 232. | 151. | 4190 |
| $D_j$ | obs | 491. | 70. | 826. | 243. | 878. | 1299. | 232. | 151. | 4190 |

Average trip length from Walsh's predicted model = 7.5827
Average trip length in SIMODEL and in observed data = 7.5836; W pred is the model prediction in Walsh S pred is the model prediction using SIMODEL

Source: Based on Walsh (1980) and authors' calculations.

## 2.4 The Alonso Framework

Alonso (1973; 1978) proposes a generalised formulation from which each of the four members of the traditional family of gravity models described above can be derived as special cases. The model proposed by Alonso can be represented by the following five equations:

$$T_{ij} = v_i A_i^{\theta-1} \, w_j B_j^{\tau-1} \, \exp(-\beta c_{ij}) \qquad (2.50)$$

$$A_i = \sum_j w_j B_j^{\tau-1} \, \exp(-\beta c_{ij}) \qquad (2.51)$$

$$B_j = \sum_i v_i A_i^{\theta-1} \, \exp(-\beta c_{ij}) \qquad (2.52)$$

$$\sum_j T_{ij} = v_i A_i^{\theta} \qquad (2.53)$$

$$\sum_i T_{ij} = w_j B_j^{\tau} \qquad (2.54)$$

where $\theta$ and $\tau$ are parameters to be specified, $v_i$ represents the propulsiveness of an origin, $w_j$ represents the attractiveness of a destination and the remainder of the notation is as defined above.

That each member of the family of spatial interaction models described above can be obtained from Alonso's framework is easily seen. Define $v_i = \sum_j T_{ij} = O_i$ and $w_j = \sum_i T_{ij} = D_j$. Then, when $\theta=1$ and $\tau=1$,

$$T_{ij} = O_i D_j \exp(-\beta c_{ij}) \qquad (2.55)$$

which is the so-called unconstrained model; when $\theta=0$ and $\tau=1$,

$$T_{ij} = O_i A_i^{-1} D_j \exp(-\beta c_{ij}) \qquad (2.56)$$

where,

$$A_i = \sum_j D_j \exp(-\beta c_{ij}) \qquad (2.57)$$

which is the production-constrained model; when $\theta=1$ and $\tau=0$,

$$T_{ij} = O_i B_j^{-1} D_j \exp(-\beta c_{ij}) \qquad (2.58)$$

where,

$$B_j = \sum_i O_i \exp(-\beta c_{ij}) \qquad (2.59)$$

which is the attraction-constrained model; and when $\theta=0$ and $\tau=0$,

$$T_{ij} = A_i^{-1} O_i B_j^{-1} D_j \exp(-\beta c_{ij}) \tag{2.60}$$

where,

$$A_i = \sum_j D_j B_j^{-1} \exp(-\beta c_{ij}) \tag{2.61}$$

and

$$B_j = \sum_i O_i A_i^{-1} \exp(-\beta c_{ij}) \tag{2.62}$$

which is the doubly constrained model.

Alonso calls $A_i$ and $B_j$ "systemic variables" although they are structurally equivalent to what are usually termed balancing factors. The parameters associated with these variables, however, relate not only inter-place flows but also total outflows and total inflows to changes within a system. This latter role occurs in equations (2.53) and (2.54) which represent trip generation and trip attraction models, respectively, and which can be derived from the first three equations by summing equation (2.50) over i or j and substituting in equation (2.52) or (2.51). An alternative interpretation of the $A_i$ and $B_j$ variables is that given by Hua (1980) who justifies their inclusion as explanatory variables in equation (2.50). By rearranging the system of equations (2.50)-(2.54), it can be shown that $A_i$ and $B_j$ are locational measures; $A_i$ is large for accessible origins where mean trip length is short and small for inaccessible origins where mean trip length is long. The variable $B_j$ can be given a similar interpretation for destinations (see Fotheringham and Dignan 1984 for more details).

Given this interpretation of $A_i$ and $B_j$ and equations (2.53) and (2.54), it might appear reasonable to interpret the parameter $\theta(\tau)$ as the elasticity of total outflow(inflow) with respect to the inverse of the mean trip length of interaction from i(to j). However, as Fotheringham and Dignan (1984) show, the values of $\theta$ and $\tau$ in equations (2.53) and (2.54) are determined primarily by the accuracy with which $v_i$ and $w_j$ represent $O_i$ and $D_j$ respectively. In equation (2.53), for example, if $v_i$ is defined as the known outflow total, then $\theta=0$ regardless of any possible relationship between $O_i$ and $A_i$. The parameters $\theta$ and $\tau$ thus indicate the relative explanatory roles of the "site" variables $v_i$ and $w_j$, and the "situation" variables $A_i$ and $B_j$. When the site variable is defined accurately, as when $v_i=O_i$, then the situation variable plays no explanatory role in the interaction model and acts purely as a balancing factor with $\theta=0$. Increasing values of $\theta$ and $\tau$ thus indicate increasingly poor definitions of the site variables $v_i$ and $w_j$, respectively. Fotheringham and Dignan's interpretation is interesting because it suggests that if there is uncertainty about the marginal totals, optimal combinations of $\theta$ and $\tau$ can exist where $0\leq\theta\leq1$ and $0\leq\tau\leq1$. That is, under some circumstances a form of quasi-constrained interaction model might be optimal. For example, a doubly-constrained model ($\theta=0$, $\tau=0$) will be optimal when total outflows and total inflows are known but in forecasting situations where there is uncertainty about these totals, the optimal combination of $\theta$ and $\tau$ may be when both parameters are non-zero.

In this way, Fotheringham and Dignan develop a family of quasiconstrained spatial interaction models from equation (2.50). Production-constrained, quasi-attraction-constrained models exist when $\theta=0$ and when $0<\tau\leq1$; attraction-constrained, quasi--production-constrained models exist when $0<\theta\leq1$ and $\tau=0$; and quasi-production-attraction constrained models exist when $0<\theta\leq1$ and $0<\tau\leq1$. The first group of models is applicable when outflow totals are known but there is uncertainty about inflow totals so that each predicted outflow total is constrained to equal $v_i$ and each inflow total is constrained to equal $w_j B_j^\tau$. The second group is applicable when inflow totals are known but there is uncertainty about outflow totals so that the predicted inflow totals are constrained to equal $w_j$ and the predicted outflow totals are constrained to equal $v_i A_i^\theta$. The third group is applicable when there is uncertainty about both the outflow and the inflow totals.

The ability of these types of model to recover known outflow and inflow totals is described in Table 2.7 which is reproduced from Table 1 in Fotheringham and Dignan (1984) and which is based on a 9 by 9 matrix of annual migration flows between the major US census divisions. The data are those given by Tobler (1983) and are discussed in more detail in section 2.6.

**Table 2.7   Examples of Quasi-Constrained Gravity Models**

| Region | $\theta=0$, $\tau=0$ Outflow Total | $\theta=0$, $\tau=0$ Inflow Total | $\theta=0$, $\tau=0.5$ Outflow Total | $\theta=0$, $\tau=0.5$ Inflow Total | $\theta=0.5$, $\tau=0$ Outflow Total | $\theta=0.5$, $\tau=0$ Inflow Total | $\theta=0.5$, $\tau=0.5$ Outflow Total | $\theta=0.5$, $\tau=0.5$ Inflow Total |
|---|---|---|---|---|---|---|---|---|
| New England | 679 | 676 | 679 | 828 | 754 | 676 | 761 | 839 |
| Middle Atlantic | 1,874 | 1,156 | 1,874 | 1,238 | 1,983 | 1,156 | 2,018 | 1,241 |
| E.N. Central | 2,134 | 1,789 | 2,134 | 1,888 | 2,196 | 1,789 | 2,198 | 1,894 |
| W.N. Central | 1,213 | 942 | 1,213 | 1,062 | 1,347 | 942 | 1,335 | 1,052 |
| S. Atlantic | 1,766 | 2,484 | 1,766 | 2,421 | 1,599 | 2,484 | 1,621 | 2,467 |
| E.S. Central | 986 | 819 | 986 | 980 | 1,180 | 819 | 1,178 | 968 |
| W.S. Central | 1,146 | 1,237 | 1,146 | 1,272 | 1,157 | 1,237 | 1,148 | 1,266 |
| Mountain | 987 | 1,067 | 987 | 1,050 | 1,003 | 1,067 | 947 | 1,000 |
| Pacific | 1,528 | 2,143 | 1,528 | 1,574 | 1,095 | 2,143 | 1,107 | 1,589 |
| Outflow or Inflow Index | 0.0 | 0.0 | 0.0 | 0.1124 | 0.1008 | 0.0 | 0.1024 | 0.1136 |

Figures are in thousands

When $\theta=0$ and $\tau=0$ the outflow and inflow totals are replicated exactly, so that these are the known marginal totals.

For each model, $v_i$ and $w_j$ are defined as the actual outflow and inflow totals so that differences between the actual and predicted totals are determined solely by the values of $\theta$ and $\tau$ and the ability to replicate the marginal totals will decrease as $\theta$ and $\tau$ diverge from 0. The differences between the actual and predicted row and column totals for each model are summarised in Table 2.7 by an outflow index and an inflow index. The outflow index is defined by

$$O = \frac{\sum_i (1 - O_i)/O_i}{n} \tag{2.63}$$

and the inflow index by

$$I = \frac{\sum_j (1 - D_j)/ D_j}{m} \qquad (2.64)$$

where n represents the number of origins and m represents the number of destinations. The indices have minimum values of zero when all the respective marginal totals are replicated exactly (in this case when $\theta=0$ and $\tau=0$ because of the definitions of $v_i$ and $w_j$) and increasing values indicate increasingly poor replication.

The results in Table 2.7 indicate the sensitivity of the differences between observed and replicated marginal totals to changes in $\theta$ and $\tau$, an issue which is explored in much greater detail by Fotheringham and Dignan (1984). They also demonstrate the obvious fact that if both row and column totals are known with certainty, then the doubly constrained model is the most appropriate member of the family of interaction models to apply. When these totals are not known with certainty, however, quasi-constrained models may be more accurate: there is little use after all in constraining a model to replicate values in which there is little confidence. Fotheringham and Dignan describe situations in which quasi-constrained spatial interaction models perform more accurately than do production--attraction models.

## 2.5 Relaxed Spatial Interaction Models

The development of a family of quasi-constrained interaction models by Fotheringham and Dignan (1984) described above within the context of Alonso's framework is similar in intent to the development of what are known as relaxed spatial interaction models. The distinguishing feature of such models is that each marginal total of the predicted interaction matrix is constrained only to be within a certain range of values, rather than being constrained to equal specific values. Relaxed models are hence useful in two situations.

(i) Where exact marginal totals for some forecast period are unknown and only estimates are available along with some idea of the error distribution associated with these estimates. For example, in a planning context a residential zone, i, might be forecast to contain 1000 ±100 dwelling units in time t so that a range of values of $O_i$ is established between 900 and 1100 with the most likely value being 1000. Similarly, an industrial zone, j, might be forecast to contain 500±5% jobs in the target year. Consequently, a range of $D_j$ values is established between 475 and 525 with the most likely value being 500.

(ii) Where marginal totals are reported independently so that there is no guarantee that the sum of the inflow totals matches the sum of the outflow totals. This inconsistency can be overcome with the use of relaxed interaction models by attributing a percentage error to each reported value so that each marginal total is constrained only to lie within a range of values.

Relaxed spatial interaction models were developed initially by Dacey and Norcliffe (1977) and Jefferson and Scott (1979). In terms of a mathematical programming problem, the relaxed interaction model corresponding to the traditional production-attraction constrained model can be written as:

$$\text{minimise } \sum_i \sum_j T_{ij} \ln T_{ij} \qquad (2.65)$$

subject to,

$$O_i^1 \le \sum_j T_{ij} \le O_i^u \quad i=1,...,m \tag{2.66}$$

$$D_j^1 \le \sum_i T_{ij} \le D_j^u \quad j=1,...,n \tag{2.67}$$

$$\sum_i \sum_j T_{ij} C_{ij} = C \tag{2.68}$$

$$\sum_i \sum_j T_{ij} = T, \quad T_{ij} \ge 0 \quad i=1,...,m; \ j=1,...,n \tag{2.69}$$

where the superscripts l and u denote the lower and upper bounds, respectively on the marginal totals. Equation (2.68) represents a trip length constraint in the usual manner.

The above programming statement makes it explicit that each predicted marginal total is allowed in theory to lie anywhere within a range of values. However, in practice, a problem often arises with this formulation. The intervals in which each marginal total is expected to lie are generally chosen so that the most likely value of each total lies close to the mid-point of the interval. Consequently, the extremes of the range, $O^1$ and $O^u$ for example in the case of outflows, may well be the least likely values expected for the predicted marginal totals. The problem that arises with the above formulation is, as noted by Hallefjord and Jornsten (1985), the solution will often lie at one of the extremes of the range: that is, the constraints are usually satisfied as equalities rather than inequalities. Thus, the model solution will often diverge significantly from the most likely predictions of the marginal totals. To alleviate this difficulty, Hallefjord and Jornsten (1985) propose the following programming formulation expressed as an information-minimisation problem:

Minimise

$$\sum_i \sum_j T_{ij} \ln T_{ij} + \delta \sum_i O_i \ln(O_i/O_i^M) + \mu \sum_j D_j \ln(D_j/D_j^M) \tag{2.70}$$

subject to:

$$\sum_j T_{ij} = O_i \quad i=1,...,m \tag{2.71}$$

$$\sum_i T_{ij} = D_j \quad j=1,...,n \tag{2.72}$$

$$\sum_i \sum_j T_{ij} C_{ij} = C \tag{2.73}$$

$$\sum_i \sum_j T_{ij} = T \quad T_{ij} \ge 0 \quad i=1,...,m; \ j=1,...,n \tag{2.74}$$

$$O_i^1 \le O_i \le O_i^u \quad i=1,...,m \tag{2.75}$$

$$D_j^1 \le D_j \le D_j^u \quad j=1,...,n \tag{2.76}$$

where the $O_i$ and $D_j$ values are now variables constrained to lie within certain ranges as defined in (2.75) and (2.76) but with the most likely values corresponding to $O_i^M$ and $D_j^M$, respectively. The modified objective function in (2.70) makes it more likely that the model's marginal totals will be close to these values since when $O_i=O_i^M$, $O_i \ln(O_i/O_i^M)=0$. Hallefjord and Jornsten interpret the parameters $\delta$ and $\mu$ as indices of the importance of meeting the target values for the row and column totals, respectively. An alternative interpretation is that they indicate the reliability of the estimates of the marginal totals and in this way act in much the same way as $\theta$ and $\tau$ in Alonso's formulation.

The values in Table 2.8 summarise the results reported in Tables 1 and 3 of Hallefjord and Jornsten (1985) and indicate the influence of the parameters $\delta$ and $\mu$ on the accuracy with which the model's marginal totals meet the <u>a priori</u> most likely values.   As $\delta$ and $\mu$ increase, the predicted marginal totals become increasingly similar to the <u>a priori</u> values. Note that there is an intentional inconsistency in the reported (most likely) sums of the outflow totals and inflow totals which is overcome in the model solutions.

Table 2.8  **Summary of Hallefjord and Jornsten's Results**

| Region | Reported Outflows | Reported Inflows | $\delta = \mu = 5.0$ | | $\delta = \mu = 50.0$ | |
|---|---|---|---|---|---|---|
| | | | Predicted Outflows | Predicted Inflows | Predicted Outflows | Predicted Inflows |
| 1 | 801 | 833 | 708 | 810 | 785 | 847 |
| 2 | 1479 | 649 | 1549 | 831 | 1462 | 685 |
| 3 | 128 | 37 | 140 | 54 | 130 | 41 |
| 4 | 146 | 171 | 196 | 254 | 153 | 186 |
| 5 | 8 | 5 | 12 | 10 | 6 | 3 |
| 6 | 262 | 94 | 231 | 104 | 261 | 101 |
| 7 | 417 | 158 | 460 | 218 | 420 | 170 |
| 8 | 1453 | 2410 | 1131 | 1932 | 1382 | 2353 |
| 9 | 531 | 411 | 533 | 438 | 529 | 428 |
| 10 | 1318 | 2067 | 1317 | 2069 | 1288 | 2072 |
| 11 | 927 | 594 | 1097 | 787 | 937 | 631 |
| 12 | 217 | 50 | 195 | 60 | 218 | 55 |
| | 7687 | 7479 | 7569 | 7567 | 7571 | 7572 |

## 2.6  The Tobler Framework

A slightly different set of spatial interaction models is suggested in a sequence of papers by Dorigo and Tobler (1983), Tobler (1983) and Ledent (1985).  Tobler's (1983) idea is alter the specification of the balancing factors from the traditional multiplicative framework to an additive framework so that in equation (2.33), for example, the term $A_iB_j$ is replaced by $A_i+B_j$.  This redefinition has implications for the calculation of balancing factors, and for the interpretation of the results, which are now discussed.  First, it can be shown that an additive formulation is the solution to a model with a quadratic objective rather than an entropy based objective.  In turn, this implies that the model produces a more dispersed set of interactions than would a linear model such as, for example, the transportation problem of linear programming.  Second, the additive formulation is also consistent with the empirically derived laws of migration postulated in Ravenstein (see Dorigo and Tobler

1983). Whatever the merits of this revised model, a useful synthesis of both the additive and multiplicative formulations is given in Ledent (1985). He shows that the Tobler additive model and the Wilson multiplicative model are at two ends of a continuum of models.

In order to illustrate the results obtained by Tobler for a migration model, it is necessary first to derive the revised set of interaction equations. Only the doubly constrained version will be discussed here for the sake of brevity. The objective is to minimise a quadratic function of interaction costs:

$$\text{minimise } W = \Sigma_i \Sigma_j T_{ij}^{2} \exp(-\beta c_{ij}) \; / \; O_i D_j \tag{2.77}$$

subject to:

$$\Sigma_j T_{ij} = O_i \, , \tag{2.78}$$

$$\Sigma_i T_{ij} = D_j \, . \tag{2.79}$$

By forming the lagrangian

$$L = W \; - \; \Sigma_i A_i (\Sigma_j T_{ij} - O_i) - \Sigma_j B_j (\Sigma_i T_{ij} - D_j) \tag{2.80}$$

and setting $\partial L / \partial T_{ij} = 0$ implies that,

$$T_{ij} = (A_i + B_j) \; O_i D_j \exp(-\beta c_{ij}) \tag{2.81}$$

where a division by 2 is subsumed in the balancing factors. Since the sum over j of the values in equation (2.81) must equal $O_i$ and the sum over i must equal $D_j$, substituting (2.81) into (2.78) and (2.79) and rearranging produces the following equations for the balancing factors $A_i$ and $B_j$, respectively:

$$A_i = \frac{1 - \Sigma_j B_j D_j \exp(-\beta c_{ij})}{\Sigma_j D_j \exp(-\beta c_{ij})} \tag{2.82}$$

$$B_j = \frac{1 - \Sigma_i A_i O \exp(-\beta c_{ij})}{\Sigma_i O_i \exp(-\beta c_{ij})} \tag{2.83}$$

This set of equations must be solved iteratively as in the traditional model (see the subsequent chapter on model calibration). To see the difference in the application of the traditional, Wilsonian type model (equations (2.40)-(2.42)) and Tobler's version (equations (2.81)-(2.83)), both models were calibrated with the US interregional migration data reported in Tobler (1983) and reproduced in full in Chapter 3. In both models costs were defined as a logarithmic function of distance so that,

$$\exp(-\beta c_{ij}) = \exp(-\beta \ln d_{ij}) = d_{ij}^{\beta} \tag{2.84}$$

where $d_{ij}$ represents the straight line distance between the centroids of regions 1 and j. Calibration of the two models (see Chapter 3) yields very similar estimates of the distance--decay parameter $\beta$. For the Wilson model $\beta=0.918$, while for Tobler's model $\beta=0.905$.

That the two models produce different predictions of interactions is demonstrated in Tables 2.9 and 2.10 in which the percentage errors from the Wilson model and the Tobler model, respectively, are presented.  Each percentage error denotes the inaccuracy of the model in replicating a particular flow.  Tobler's model produces more accurate predictions than Wilson's model for 45 of the 72 interactions.  However, many of the differences are extremely small as one might expect given the similarity of the two formulations.  What is intriguing is the reason for the obvious superiority of Tobler's model in some instances. In particular, the flows between the W N Central region and the New England, E S Central, and Pacific regions, between New England and E S Central, and between the Pacific and S Atlantic regions are replicated substantially more accurately by Tobler's model.

**Table 2.9  Percentage Errors from a Traditional Production-Attraction Constrained Interaction Model in Replicating US Interregional Migration**

|                 | 1     | 2     | 3     | 4     | 5     | 6     | 7     | 8     | 9     |
|-----------------|-------|-------|-------|-------|-------|-------|-------|-------|-------|
| 1 New England   | –     | 29.0  | 16.9  | 10.7  | -23.1 | 50.6  | 10.4  | -7.9  | -29.9 |
| 2 Mid Atlantic  | 11.7  | –     | 9.3   | 47.3  | -20.6 | 73.7  | 40.6  | 1.8   | -9.3  |
| 3 E-N Central   | 9.6   | 5.0   | –     | 2.3   | -4.9  | -24.2 | 42.6  | -7.3  | -1.8  |
| 4 W-N Central   | 9.2   | 31.6  | 6.7   | –     | 37.5  | 32.5  | -6.8  | -36.1 | -12.1 |
| 5 S Atlantic    | -20.4 | -25.1 | 0.2   | 33.5  | –     | 2.4   | 16.2  | 25.7  | 15.4  |
| 6 E-S Central   | 27.2  | 31.6  | -40.7 | 21.7  | 14.4  | –     | -14.1 | 69.0  | 44.1  |
| 7 W-S Central   | 3.5   | 19.2  | 21.7  | -13.2 | 30.7  | -20.4 | –     | -26.6 | -10.4 |
| 8 Mountain      | -4.3  | 9.9   | 15.7  | -32.0 | 33.6  | 31.5  | -43.0 | –     | 10.8  |
| 9 Pacific       | -32.5 | -15.8 | 2.2   | -16.6 | 11.1  | 19.0  | -19.1 | 21.3  | –     |

Negative values indicate underprediction by the model

**Table 2.10   Percentage Errors from Tobler's Production-Attraction Constrained Interaction Model in Replicating US Interregional Migration**

|                 | 1     | 2     | 3     | 4     | 5     | 6     | 7     | 8     | 9     |
|-----------------|-------|-------|-------|-------|-------|-------|-------|-------|-------|
| 1 New England   | –     | 28.8  | 15.6  | 4.3   | -22.3 | 35.5  | 9.3   | -6.2  | -26.2 |
| 2 Mid Atlantic  | 9.7   | –     | 9.6   | 42.7  | -19.6 | 63.5  | 40.5  | 3.4   | 7.7   |
| 3 E-N Central   | 8.4   | 5.2   | –     | 1.8   | -4.4  | -26.3 | 43.9  | -5.4  | -2.3  |
| 4 W-N Central   | -1.6  | 27.1  | 5.6   | –     | 37.4  | 17.6  | -7.3  | -34.6 | -6.6  |
| 5 S Atlantic    | -14.6 | -23.6 | 1.6   | 36.9  | –     | 2.9   | 17.0  | 25.0  | 2.8   |
| 6 E-S Central   | 4.7   | 24.1  | -42.7 | 8.2   | 16.4  | –     | -15.7 | 72.3  | 62.6  |
| 7 W-S Central   | 3.0   | 19.0  | 22.9  | -13.1 | 31.2  | -21.6 | –     | -25.0 | -11.6 |
| 8 Mountain      | -5.0  | 9.4   | 16.4  | -32.1 | 33.1  | 27.9  | -42.4 | –     | 10.9  |
| 9 Pacific       | -22.2 | -14.7 | 3.0   | -10.7 | 4.2   | 33.9  | -19.6 | 18.1  | –     |

## 2.7 Summary

This chapter has reviewed a wide variety of spatial interaction models and positioned them against a background of optimisation tasks. Wilson's original derivation of these models using entropy techniques opened the way for more general formulations, such as those based on prior information (Snickars and Weibull, 1977); those based on a paramaterised set of accounting identities (Alonso, 1978; Fotheringham and Dignan, 1984); problems with inequality constraints rather than strict equalities (Hallefjord and Jornsten, 1985); and finally those problems with hybrid objective functions (Tobler, 1983). The remainder of the book now focuses attention on formulating, calibrating and applying members of this extended family of models.

# CHAPTER 3

---

## THE CALIBRATION OF SPATIAL INTERACTION MODELS

### 3.1 Introduction

In the previous chapter, four specific forms of spatial interaction model were derived from the general formulation in equation (1.7). These were the unconstrained, production-constrained, attraction-constrained and doubly constrained models. Common to each is the need to obtain accurate estimates of the model's parameters. This is necessary in order to forecast interactions in different systems and it is also useful in providing information on the system under investigation (see Chapter 1). The process of obtaining estimates of a model's parameters is known as model calibration.

To calibrate an interaction model it is necessary to have a known interaction matrix. The question that might be asked is "if we already have the interaction matrix, why do we need to calibrate an interaction model?" There are three major reasons why we might want to undertake such an operation:

(i)     To predict interactions at some future time period or in a different spatial system.

(ii)    To forecast the effects on interaction patterns of planned changes in spatial structure. For example, if the interaction data consisted of journeys-to-work in an urban area, it would be possible to forecast the changes in traffic patterns that would result if a major industrial development took place in some part of the city or if a new road were built.

(ii)    Model calibration yields parameter estimates that can provide information on the system under investigation and it is possible to draw conclusions about interaction behaviour from a comparison of parameter estimates. For example, if a model were calibrated in the same spatial system with migration data from two time periods, it might be possible to conclude, from a comparison of the two distance-decay parameters, that migrants are becoming less or more constrained by distance over time. The research task would then focus on explaining why this change is occurring.

This chapter discusses techniques that can be employed to obtain estimates of the parameters of the four interaction models described in Chapter 2. Because of the similarity of the production-constrained and attraction-constrained models, however, only the former model is discussed. Two different methods of calibrating each model are outlined--ordinary least squares regression (OLS) and maximum likelihood estimation (MLE). It is not the task of this book to discuss these techniques in detail (Lewis-Beck (1980) and Achen (1982) provide introductions to regression while Hanushek and Jackson (1977) provide a useful description of maximum likelihood estimation); rather, we concentrate on the application of these techniques to the calibration of spatial interaction models.

### 3.2 The Calibration of Spatial Interaction Models by Regression

In order to be calibrated by regression, a model must be in a linear format, that is, linear in terms of its parameters. Hence, the models described in equations (2.20), (2.21), (2.26) and (2.33) must be transformed into the following format:

$$y = \alpha_0 + \alpha_1 x_1 + \alpha_2 x_2 + \dots \alpha_n x_n \tag{3.1}$$

where, in standard regression terminology, $y$ is the dependent variable (a function of interaction in this case), the $x$'s represent independent variables (functions of the $v_i$s, $w_j$s and $c_{ij}$s), $\alpha_0$ is a scale parameter and the remaining $\alpha$s represent the relationship between particular independent variables and the dependent variable. The necessary linear transformation is now described for each of the spatial interaction models.

### 3.2.1   The Unconstrained Model

The transformation of the unconstrained model,

$$T_{ij} = k v_i^{\mu} w_j^{\alpha} c_{ij}^{\beta} \tag{3.2}$$

into a linear format is achieved very easily by simply taking logarithms of both sides of the equation so that equation (3.2) becomes:

$$\ln T_{ij} = \ln k + \mu \ln v_i + \alpha \ln w_j - \beta \ln c_{ij} \tag{3.3}$$

where ln denotes a natural logarithm. Note that purely for pedagogic purposes, the simplest form of the model is represented. The addition of other variables would not alter the basic transformations that are the focus of this discussion. Where the functional forms of the variables in each model are made explicit, as above, only the transformations using power functions are reported in the interests of brevity. For instance, if equation (3.2) contained an exponential cost function, the linear transformation would clearly be

$$\ln T_{ij} = \ln k + \mu \ln v_i + \alpha \ln w_j - \beta c_{ij} . \tag{3.4}$$

If the assumptions of OLS regression are met, $\mu'$, $\alpha'$ and $\beta'$ are unbiased and consistent estimates of $\mu$, $\alpha$ and $\beta$, respectively, but $\exp(\ln k)'$ is a biased estimate of $k$ [see Heien (1968) and Haworth and Vincent (1979)]. In fact, $k$ will always be underestimated when obtained by OLS unless the model fit is perfect. The underestimation of $k$ results in

$$\sum_i \sum_j T_{ij}' \leq \sum_i \sum_j T_{ij} \tag{3.5}$$

so that a more accurate estimate of $k$, $k'$(new), should be obtained after the regression in the following manner:

$$k'(\text{new}) = k'(\text{old}). \sum_i \sum_j T_{ij} / \sum_i \sum_j T_{ij}' . \tag{3.6}$$

A simple example of the underestimation of total flows that can occur is given below in section 3.3 where this important, but often ignored, feature of logarithmic regression is discussed further along with two other problems that can arise in the regression-based calibration of spatial interaction models.

Potential heteroscedacity problems can occur in the calibration of equation (3.3) by regression due to the logarithmic transformation (Stronge and Schultz, 1978) although this problem can usually be solved by calibrating the model by weighted or generalised least squares regression, with a weight equal to $(\ln T_{ij})^{1/2}$, rather than by OLS regression. The OLS estimators are unbiased and consistent but generally have larger variances (that is, are less efficient) than GLS estimators so that use of the former is less likely to detect significant relationships. Empirical research reported by van Est and van Setten (1979) and by Nakanishi and Cooper (1975), however, questions whether there is any practical difference between the two techniques in terms of parameter estimation. Useful descriptions of the heteroscedacity problem and weighted least squares regressions are given by

Hanushek and Jackson (1977), Gujarati (1978) and Cooper and Weekes (1983).

## 3.2.2   The Production-Constrained (Attraction-Constrained) Model

Consider the following general form of the production-constrained interaction model:

$$T_{ij} = O_i \exp[\Sigma_h \alpha_h f_h(x_{ijh})]/\Sigma_k \exp[\Sigma_h \alpha_h f_h(x_{ikh})] \tag{3.7}$$

where $x_{ijh}$ is the hth explanatory variable and $f_h$ is the functional form of that variable in the model. The transformation of this general structure (equivalent to that found in the attraction-constrained model) into a form linear in parameters was first described for a specific model by Nakanishi and Cooper (1974) and has been used by Stetzer (1976) and van Est and van Setten (1979) and others. To understand the transformation, multiply together the set of flows emanating from each origin to the n destinations so that

$$\Pi_j T_{ij} = O_i^n \Pi_j \exp[\Sigma_h \alpha_h f_h(x_{ijh})]/(\Sigma_k \exp[\Sigma_h \alpha_h f_h(x_{ikh})])^n . \tag{3.8}$$

Take the nth root of both sides of the equation,

$$(\Pi_j T_{ij})^{1/n} = \frac{O_i \{\Pi_j \exp[\Sigma_h \alpha_h f_h(x_{ijh})]\}^{1/n}}{\Sigma_k \exp[\Sigma_h \alpha_h f_h(x_{ikh})]} . \tag{3.9}$$

Divide both sides of the equation into $T_{ij}$ and substitute for $T_{ij}$ in the right-hand side:

$$T_{ij}/(\Pi_j T_{ij})^{1/n} = \exp[\Sigma_h \alpha_h f_h(x_{ijh})]/\{\Pi_j \exp[\Sigma_h \alpha_h f_h(x_{ijh})]\}^{1/n} \tag{3.10}$$

and then by taking logarithms of both sides and rearranging, this equation is made linear in terms of its parameters

$$\ln T_{ij} - (1/n)\Sigma_j \ln T_{ij} = \Sigma_h \alpha_h [f_h(x_{ijh}) - (1/n)\Sigma_j f_h(x_{ijh})] . \tag{3.11}$$

Note that if $f_h$ is a unitary function, so producing an exponential function in the original model, then the expression $(1/n)\Sigma_j f_h(x_{ijh})$ is an arithmetic mean. If $f_h$ is a logarithmic function, so producing a power function in the original model, the expression is a geometric mean.

While equation (3.11) is the form of the production-constrained model that is most often calibrated by regression, it can be rearranged to the following:

$$\ln T_{ij} = \Sigma_h \alpha_h [f_h(x_{ijh})] + (1/n)\Sigma_j \ln T_{ij} - (1/n)\Sigma_h \alpha_h \Sigma_j f_h(x_{ijh}) \tag{3.12}$$

which can be simplified to:

$$\ln T_{ij} = k_i + \Sigma_h \alpha_h [f_h(x_{ijh})] \tag{3.13}$$

which is merely an unconstrained model with an origin-specific constant term. Cesario (1975b) was amongst the first to note this relationship.

Apart from allowing the production-constrained model (and hence the attraction-constrained model) to be calibrated by regression, the linearising technique described above has another useful property. It can be used to obtain estimates of either origin propulsiveness or destination attractiveness. For example, suppose the patronage of state parks is of interest and the number of people visiting the parks from various places is known, but an accurate measure of the attractiveness of each park is not available (a recurring problem in recreation and tourism studies). Simply using the number of visitors to each park as a measure of attractiveness would be misleading since some parks will obviously be closer to larger centers of population than others. A measure of attractiveness is needed that is independent of location and this can be provided as follows. Let the unknown attractiveness of a destination be denoted by $a_j$ and let spatial separation be represented as a power function of distance. The production-constrained model can than be written as:

$$T_{ij} = O_i a_j d_{ij}^{-\beta} / \sum_k a_k d_{ik}^{-\beta} \qquad (3.14)$$

which, in linearised form, becomes,

$$\ln T_{ij} - (1/n)\sum_j \ln T_{ij} =$$

$$(\ln a_j - (1/n)\sum_j \ln a_j) - \beta(\ln d_{ij} - (1/n)\sum_j \ln d_{ij}) . \qquad (3.15)$$

When equation 3.15 is calibrated by OLS, the constant term in the regression is an estimate of $(\ln a_j - (1/n)\sum_j \ln a_j)$ which can be used to yield values of the unknown $a_j$s. Dummy variables must be used in order to obtain estimates of all the $a_j$s and, to avoid perfect multicollinearity, one of the destinations must be excluded from the regression. The attractiveness of that destination is then obtained from the constant term alone. For the other destinations, the attractiveness is derived from the constant term plus the relevant dummy variable parameter. If the attractiveness of the excluded destination is defined as some constant, say 1.0, then this defines $(1/n)\sum_j \ln a_j$ and all of the other attractiveness values can then be determined relative to the one  set at 1.0. In essence, this technique is akin to estimating a regression model with an origin-specific constant term [Cesario (1975b)]. Baxter and Ewing (1981) provide empirical examples of the estimation of attractiveness in this way and an example is provided below in section 3.6.

### 3.2.3  The Doubly Constrained Model

At first glance it might appear impossible to linearise the doubly constrained model from section 2.3.4 written here with a power function of distance:

$$T_{ij} = A_i O_i B_j D_j d_{ij}^{-\beta} \qquad (3.16)$$

where,

$$A_i = 1/\sum_j B_j D_j d_{ij}^{-\beta} \qquad (3.17)$$

and

$$B_j = 1/\sum_i A_i O_i d_{ij}^{-\beta} . \qquad (3.18)$$

However, two techniques have recently been described that achieve this task. Sen and Soot (1981) and Gray and Sen (1983) have provided a technique that separates the estimation of the distance (cost) parameter from the calculation of balancing factors. The procedure is termed the odds ratio method and involves taking ratios of interactions so that the $A_i O_i$ and $B_j D_j$ terms in the model cancel out. From equation (3.16),

$$T_{ij} = A_i O_i B_j D_j d_{ij}^{-\beta} \qquad (3.19)$$

$$T_{ji} = A_j O_j B_i D_i d_{ji}^{-\beta} \qquad (3.20)$$

$$T_{ii} = A_i O_i B_i D_i d_{ii}^{-\beta} \qquad (3.21)$$

$$T_{jj} = A_j O_j B_j D_j d_{jj}^{-\beta} \qquad (3.22)$$

so that

$$(T_{ij}/T_{ii}).(T_{ji}/T_{jj}) = [(d_{ij}/d_{ii}).(d_{ji}/d_{jj})]^{-\beta} . \tag{3.23}$$

The equation is then made linear in its parameters by taking logarithms,

$$\ln T_{ij} + \ln T_{ji} - \ln T_{ii} - \ln T_{jj} =$$

$$-\beta(\ln d_{ij} + \ln d_{ji} - \ln d_{ii} - \ln d_{jj}) \tag{3.24}$$

Sen and Soot suggest using weighted least squares to counteract the heteroscedastic error terms caused by the logarithmic transformation with the weight being $(T_{ij}^{-1} + T_{ji}^{-1} + T_{ii}^{-1} + T_{jj}^{-1})^{-0.5}$ .

Once $\beta$ is estimated, the balancing factors of the model can be obtained by iterating equations (3.17) and (3.18).

Two potential problems can arise with the above linear transform of the doubly-constrained model. The interaction matrix has to be square and it is necessary for the intrazonal interactions ($T_{ii}$ and $T_{jj}$) to be non-zero. An alternative method of linearising the doubly-constrained gravity model that does not suffer from these problems is given by Sen and Soot (1981) and uses the relationship that,

$$\ln T_{ij} - (1/n)\sum_j \ln T_{ij} - (1/m)\sum_i \ln T_{ij} + (1/mn)\sum_i\sum_j \ln T_{ij}$$

$$= -\beta \ [\ln d_{ij} - (1/n)\sum_j \ln d_{ij} - (1/m)\sum_i \ln d_{ij} + (1/mn)\sum_i\sum_j \ln d_{ij}] \tag{3.25}$$

where $(1/n)\sum_j \ln T_{ij}$ is the row mean of the $\ln T_{ij}$s; $(1/m)\sum_i \ln T_{ij}$ is the column mean; and $(1/mn)\sum_i\sum_j \ln T_{ij}$ is the grand mean.

Again, weighted least squares may be preferable to ordinary least squares. In this instance, the weight is simpler, being $T_{ij}^{0.5}$. This linearisation technique can be used for rectangular matrices. However, it needs to be modified when intrazonal flows are zero (Sen and Pruthi, 1983).

## 3.3  Cautionary Notes on the Calibration of Interaction Models by Regression

The transformations described above provide a useful mechanism for calibrating interaction models by regression which has the advantage of being a readily available calibration technique with well-known properties. There are, however, three potential problems that can arise with the technique. The first concerns the estimation of the constant term in an unconstrained model, which as already discussed, tends to be biased downwards; the second concerns goodness-of-fit calculations; and the third concerns the replacement of zero flows. Each is now examined in more detail.

### 3.3.1  Bias in the Constant Term

The constraint operating in regression is that,

$$\sum_i\sum_j e_{ij} = 0 \tag{3.26}$$

where $e_{ij}$ represents the error term in the regression.  This implies that

$$\sum_i\sum_j y_{ij}' = \sum_i\sum_j y_{ij} \tag{3.27}$$

where $y_{ij}$ represents the transformed interaction variable in one of the transformations described above and $y_{ij}'$ is the predicted value of $y_{ij}$. In the case of the unconstrained model, for example,

$$\sum_i\sum_j \ln T_{ij}' = \sum_i\sum_j \ln T_{ij} \qquad (3.28)$$

which does not imply that the desired constraint,

$$\sum_i\sum_j T_{ij}' = \sum_i\sum_j T_{ij} \qquad (3.29)$$

is met. In fact, the constraint in (3.28) ensures that, unless the model is perfectly accurate,

$$\sum_i\sum_j T_{ij}' < \sum_i\sum_j T_{ij} \qquad (3.30)$$

that is, the total flow volume predicted will be less than the actual total flow volume. If the model is imperfect, the variance of the predicted values will be less than the variance of the actual values. Consequently, there is a tendency for small flows to be overpredicted and large flows to be underpredicted. While these over- and underpredictions cancel each other out in logarithms, in terms of real flows, the underpredictions of large flows will be greater than the overprediction of small flows. Hence, the discrepancy described in (3.30) arises.

This can be demonstrated in an alternative way. The average predicted value from the model, $\exp\{[1/(mn)]\sum_i\sum_j\ln T_{ij}'\}$, is actually the geometric mean of the $T_{ij}'$ values. By a proof given by Beckenbach and Bellman (1961), the geometric mean of a set of values is always less than or equal to the arithmetic mean. Hence,

$$\sum_i\sum_j \exp(\ln T_{ij}') \le \sum_i\sum_j T_{ij} \; . \qquad (3.31)$$

To remove this discrepancy, it is necessary to revise the estimate of the constant term in the model through the use of equation (3.6) or, as Heien (1968) and Haworth and Vincent (1979) suggest, by performing the following transformation on the exponential of the estimated constant from a logarithmic regression:

$$k'(\text{new}) = k'(\text{old}).\exp(\hat{s}/2) \qquad (3.32)$$

where $s^2$ is the sample variance of the logarithmic error terms from the regression. Since $s^2 \ge 0$, $\exp(s^2/2) \ge 1$ and $k'(\text{new}) \ge k'(\text{old})$.

### 3.3.2   Goodness-of-fit Calculations and Logarithmic Regressions

Generally, the calibration of models by regression is undertaken with statistical computer software such as SPSS-X, SAS, and the various micro-based packages. It is important to realise that the goodness-of-fit statistic(s) reported in such packages, which are generally based on an R-squared measure, relate the accuracy of the model in replicating the dependent variable; the latter not being interaction but one of the transformations of interaction described above. Thus, the goodness-of-fit statistics reported in such packages can be misleading: we are interested in predicting interaction, not some transformation of it. It is therefore necessary to transform the predicted dependent variable into a predicted interaction and then calculate a goodness-of-fit statistic with those predicted interactions. A discussion on goodness-of-fit statistics suitable for assessing the performance of interaction models is given below in section 3.7.

### 3.3.3   The Problem of Zero Interactions

Because all of the transformations described above involve taking a logarithm of interaction, a problem arises whenever the sampled interaction between any two points is zero since the logarithm is then undefined. Several solutions to this problem, of varying degrees of satisfaction, are possible. Perhaps the most obvious solution is simply to remove all zero interactions from the analysis. However, the resulting parameter estimates would not reflect the low volumes of interaction that occur between certain origins and destinations and so would be misleading. A second solution is to remove from the analysis all origins and destinations associated with zero interactions. However, a great deal of useful information can be lost in this way, and in particularly sparse matrices, there may be no origin that has a non-zero interaction to every destination.

The third solution is by far the most commonly used in dealing with zero interactions and it involves adding a constant to elements of the interaction matrix. Two possibilities exist here: one is to add the constant to every flow in the matrix; the other is to add the constant only to the zero flows. In practice, there seems little difference between the two methods in terms of the resulting parameter estimates. In both cases, some uncertainty exists over the value of the constant to be added. Probably the most frequently encountered method of dealing with zero interactions is to add one to every zero flow and this can be justified on the grounds that the flows recorded are generally integer and one is the closest approximation to zero. It also ensures that the logarithmic interactions have a minimum of zero. However, Sen and Soot (1981) provide a theoretical justification for the addition of 0.5 rather than one although their justification does not take into account possible effects on parameter estimates.

None of the above problems is encountered in the second method of calibration, maximum likelihood estimation (MLE), which is now discussed. There is a trade-off, however, in that MLE calibration routines tend to be less accessible than their regression counterparts but in section 3.8 a computer programme developed especially for the calibration of spatial interaction models by maximum likelihood is discussed.

### 3.4   The Calibration of Models by Maximum Likelihood Estimation

In essence, the technique of MLE is to find parameter estimates that maximise the likelihood of observing a sample set of interactions from a theoretical distribution. The steps involved in the calibration include identifying a theoretical distribution for the interactions, maximising the likelihood function of this distribution with respect to the parameters of the interaction model, and then deriving equations that ensure the maximisation of the likelihood function. For convenience, the logarithm of the likelihood function is usually used since this is at a maximum whenever the likelihood function is at a maximum (see Pickles, 1986, for a demonstration of this). Parameter estimates that maximise the likelihood function are termed maximum likelihood estimates. Maximum likelihood estimators have several desirable properties: they are consistent, asymptotically efficient and are asymptotically normally distributed. This latter property is particularly useful in significance testing and further comment on this subject is made below.

The method of obtaining parameter estimates by MLE is described for each of the four interaction models although again the method is identical for the production-constrained and attraction-constrained models so is only discussed in terms of the former.

### 3.4.1   The Unconstrained Model

Flowerdew and Aitkin (1982) state that interactions can be considered to be the outcome of a Poisson process if it is assumed that there is a constant probability of any individual in i moving to j, that the population of i is large, and that the number of individuals interacting is an independent process. Consequently, the probability that $T_{ij}$ is the number of people recorded as moving between i and j is given by,

$$p(T_{ij}) = [\exp(-T_{ij}').T_{ij}'^{(T_{ij})}]/T_{ij}! \tag{3.33}$$

where $T_{ij}'$ is the expected outcome of the Poisson process. Note that this is distinguished from the observed value $T_{ij}$, the latter being subject to sampling and measurement errors and therefore fluctuates around the expected value, $T_{ij}'$. Since $T_{ij}'$ is unknown and unobservable, it has to be estimated from some theoretical model such as the unconstrained model in equation (3.2).

Consider the log-likelihood of a set of observed flows $\{T_{ij}\}$ where each flow is the outcome of a particular Poisson process. This log-likelihood, L*, can be represented as:

$$L^* = \Sigma_i\Sigma_j \ln[\exp(-T_{ij}').T_{ij}'^{(T_{ij})}]/T_{ij}! \tag{3.34}$$

which is equivalent to

$$L^* = \Sigma_i\Sigma_j(-T_{ij}' + T_{ij}\ln T_{ij}' - \ln T_{ij}!) . \tag{3.35}$$

Since $T_{ij}$ is given, $\ln T_{ij}!$ can be ignored in the maximisation and L* will be a maximum when

$$Z = \Sigma_i\Sigma_j(T_{ij}\ln T_{ij}' - T_{ij}') \tag{3.36}$$

is at a maximum. Hence, the parameter estimates associated with $T_{ij}'$ that maximise Z are required. These are the estimates of the parameters in equation (3.2) that maximise the expression for Z in equation (3.36). Using calculus, these estimates are obtained when

$$\partial Z/\partial \xi = \Sigma_i\Sigma_j T_{ij}'\ln x_{ij} - \Sigma_i\Sigma_j T_{ij}\ln x_{ij} = 0 \tag{3.37}$$

where $x_{ij}$ is an independent variable in equation (3.2) and $\xi$ is the parameter associated with that variable. For example, if k in equation (3.2) is defined as $e^\mu$, the constraint equation for $\mu$ is

$$\partial Z/\partial \mu = \Sigma_i\Sigma_j T_{ij}' - \Sigma_i\Sigma_j T_{ij} = 0 \tag{3.38}$$

since e, the variable associated with $\mu$ is a constant. Equation (3.38) indicates that the sum of the predicted interactions ($\Sigma_i\Sigma_j T_{ij}'$) will be equal to the sum of the observed interactions ($\Sigma_i\Sigma_j T_{ij}$). This means that a total flow constraint is met automatically in the maximum likelihood calibration of the unconstrained gravity model.

In a similar manner, the maximum-likelihood equation for $\beta$, the distance-decay parameter, will be (from 3.37):

$$\Sigma_i\Sigma_j T_{ij}'\ln d_j - \Sigma_i\Sigma_j T_{ij}\ln d_{ij} = 0 \tag{3.39}$$

which can be interpreted as a cost constraint.

3.4.2 The Production-Constrained (Attraction-Constrained) Model

Interactions can be assumed to have a multinomial distribution (Batty and Mackie, 1972), in which case the log-likelihood of observing a set of flows is

$$L^{*} = \sum_i\sum_j T_{ij}\ln p_{ij} \tag{3.40}$$

where $p_{ij}$ represents the predicted probability of moving between i and j and is defined as

$$p_{ij} = T_{ij}'/\sum_i\sum_j T_{ij}' . \tag{3.41}$$

Alternatively, $p_{ij}$ can be defined as the product of two other probabilities:

$$p_{ij} = p_{j|i}\cdot p_i \tag{3.42}$$

where $p_{j|i}$ is the conditional probability of interacting with j given one originates at i, and $p_i$ is the probability of an interaction originating in i. In a production-constrained model,

$$\sum_j p_{j|i} = 1 \qquad \text{for all } i \tag{3.43}$$

and, clearly,

$$\sum_i p_i = 1 \tag{3.44}$$

The probability $p_{j|i}$ is given by a production-constrained model,

$$p_{j|i} = A_i \exp[\sum_h \alpha_h f_h(x_{ijh})] \tag{3.45}$$

so that the objective is to determine the estimates of the $\alpha_h$s which maximise equation (3.40) subject to the constraints in equations (3.43) and (3.44). It is relatively straightforward to demonstrate that the solution to this is a series of equations of the form:

$$\sum_i\sum_j T_{ij}'f(x_{ijh}) = \sum_i\sum_j T_{ij}f(x_{ijh}) \qquad \text{for all } h . \tag{3.46}$$

In each case the MLE of $\alpha_h$ is therefore obtained when the constraint in equation (3.46) is met. For example, in the case of a power distance function, the estimate of $\beta$ is obtained when

$$\sum_i\sum_j T_{ij}'\ln d_j = \sum_i\sum_j T_{ij}\ln d_{ij} \tag{3.47}$$

which is the same constraint as in the maximum-likelihood calibration of the unconstrained model. The only difference in the two models is that the latter has only a single constant to be estimated whereas the production-constrained model has a separate constant, $A_i$, to be estimated for each origin. The estimate of $A_i$ is obtained when

$$\sum_j T_{ij}' = \sum_j T_{ij} \qquad \text{for all } i . \tag{3.48}$$

### 3.4.3  The Doubly Constrained Model

The maximum likelihood estimator of the cost parameter in the doubly constrained model is obtained through the same form of constraint equation as given in (3.46) above. The only difference in the calibration of this model from that of the production-constrained model being that an extra set of parameters, the $B_j$s, is estimated from the destination constraint set

$$\sum_i T_{ij}' = \sum_i T_{ij} \qquad \text{for all } j . \tag{3.49}$$

## 3.5   An Algorithm for Maximum-Likelihood Calibration

As a demonstration of the general maximum-likelihood calibration of an interaction model, consider the estimation of $\beta$, the distance(cost)-decay parameter, in a doubly constrained model with a power function of distance. The constraint equation is that given in (3.47) and the general procedure for obtaining an estimate of $\beta$ using this equation is outlined in Figure 3.1. Starting values for $\beta$' (usually 1.0 in a negative power function) and the set of $B_j$ values (usually 1.0 for each $B_j$) are chosen. The initial values of $A_i$ are then calculated and are input into the updated calculation of the $B_j$s. This iterative calculation of the balancing factors continues until all the values are stable under further iteration. The stable $A_i$ and $B_j$ values are then input into the model and a set of predicted interactions is obtained (with $\beta = 1.0$). If the constraint equation is met at this stage, the value of $\beta$' is retained; if it is not, then $\beta$' is changed and the whole cycle repeated until the constraint equation for $\beta$ is met.

Variations in the above procedure only occur in the way in which $\beta$' is changed and algorithms of varying degrees of speed and complexity exist for this purpose. The simplest and slowest, for example, is a straightforward iteration where $\beta$' is changed by a set amount on each iteration. Other, faster procedures include first order iteration, Newton-Raphson techniques and the Secant method on which further discussion can be found in Batty and Mackie (1972), Batty (1976a) and Cheney and Kincaid (1980). Here, we will briefly outline Newton's method because it is probably the most widely applied procedure for solving a set of nonlinear equations and because it forms the basis of the SIMODEL algorithm which is discussed below.

### 3.5.1   Newton's Method for Solving One Nonlinear Equation

For any given model, the maximum-likelihood constraint equation for each parameter in that model can be represented in a general form by,

$$f(\beta) = 0 \tag{3.50}$$

where $\beta$ represents the parameter to be estimated in the model. The calibration of the model takes place by finding the value of $\beta$ generating $f(\beta) = 0$ and this is denoted by $\beta$' in Figure 3.2. The specific form of $f(\beta)$ is the nonlinear equation

$$\sum_i\sum_j T_{ij}'f(x_{ij}) - \sum_i\sum_j T_{ij}f(x_{ij}) = 0 \tag{3.51}$$

where $T_{ij}$' is a function of $\beta$. The function is differentiable over all $\beta$ and so the graph of $f(\beta)$ against $\beta$ has a definite slope at each point and hence has a unique tangent at that point. Consider the tangent to the curve at the point $\beta_0$, $f(\beta_0)$, denoted by the line $P\beta_1$ in Figure 3.2. From simple trigonometry, the line $P\beta_1$ at $\beta_0$ is defined in terms of the linear function,

$$L(\beta) = f'(\beta_0)(\beta - \beta_0) + f(\beta_0) \tag{3.52}$$

Figure 3.1  **Maximum Likelihood Calibration Procedure**

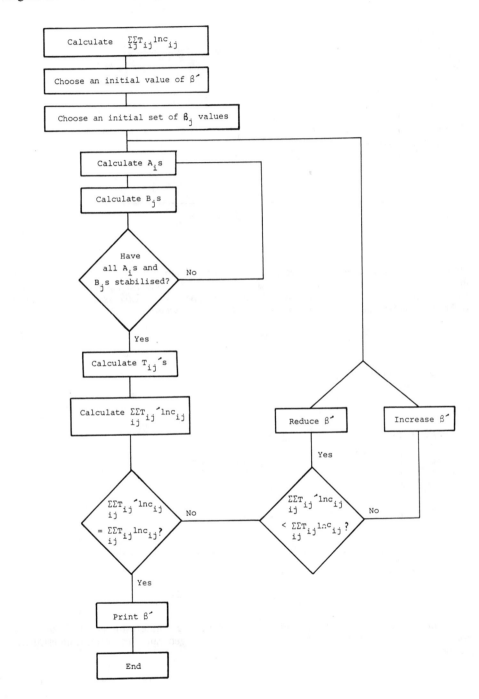

Figure 3.2  **Graph of f($\beta$) against $\beta$**

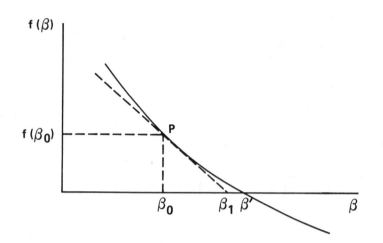

where $f'(\beta_0)$ is the slope of the curve at $\beta_0$ (and so is the slope of the tangent to this point) and $f(\beta_0)$ is the value of $f(\beta)$ at $\beta_0$. Thus, as $\beta$ increases beyond $\beta_0$, the value of $L(\beta)$ decreases by $f'(\beta)(\beta - \beta_0)$, as shown in Figure 3.3. The zero of $L(\beta)$ is an approximation to the zero of $f(\beta)$ and from equation (3.52) this occurs when

$$\beta_1 = \beta_0 - [f(\beta_0)/f'(\beta_0)] . \qquad (3.53)$$

Figure 3.3  **Determining $\beta_1$**

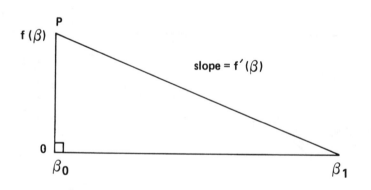

Thus, starting with a point $\beta_0$ (an initial guess), a more accurate estimate, $\beta_1$, is obtained from the formula in equation (3.53). This can be used to generate a more accurate estimate, $\beta_2$, by

$$\beta_2 = \beta_1 - [f(\beta_1)/f'(\beta_1)] \qquad (3.54)$$

and so on until $\beta$ converges. In general, the formula that is iterated is,

$$\beta_{n+1} = \beta_n - [f(\beta_n)/f'(\beta_n)] . \tag{3.55}$$

The function $f'(\beta)$ is simply the derivative of the nonlinear function with respect to the parameter being estimated.

### 3.5.2 Newton's Method for Solving a Set of Nonlinear Equations

The following discussion is based largely on that of Johnston (1982) in which an extension of the previous one-parameter estimation procedure is made so that it can be used to estimate any number of parameters.

Single parameter estimation deals with a one-dimensional curve as in Figure 3.2. For n parameters it is necessary to deal with an n-dimensional hypersurface. The derivative $f'(\beta)$ at $\beta = \beta_k$ is a number $f'(\beta_k)$ that satisfies the condition

$$[f(\beta_k) - f(\beta_p)]/(\beta_k - \beta_p) - f'(\beta_k) \ \to 0 \text{ as } \beta_p \to \beta_k \tag{3.56}$$

which is equivalent to

$$f(\beta_k) - f(\beta_p) - f'(\beta_k).(\beta_k - \beta_p) \ \to 0 \text{ as } \beta_p \to \beta_k. \tag{3.57}$$

The analogue of $f'(\beta_k)$ in n dimensions is an n x n matrix, $J(\beta_k)$, that satisfies the condition

$$f(\beta_k) - f(\beta_p) - J(\beta_k).(\beta_k - \beta_p) \ \to 0 \text{ as } \beta_p \to \beta_k \tag{3.58}$$

where $J(\beta_k)$ is a Jacobian matrix whose entries are the partial derivatives,

$$J_{ab} = \partial f_a(\beta)/\partial \beta_b \tag{3.59}$$

where the indices a and b refer to parameters to be estimated. Then, given a set of initial estimates of $\beta$, a more accurate set of estimates, $\beta_k$, is given by

$$\beta_{k+1} = \beta_k - [J(\beta_k)]^{-1}.f(\beta_k). \tag{3.60}$$

Computationally, the above procedure involves the following steps;

(i)   Compute $f(\beta_k)$ and $J(\beta_k)$
(ii)  Invert $J(\beta_k)$
(iii) Compute $\beta_k - [J(\beta_k)]^{-1}.f(\beta_k)$
(iv)  Using the value in (iii) as k return to (i)

### 3.5.3 Standard Errors of Maximum Likelihood Parameter Estimates

Consistent estimators of the asymptotic variances of the parameter estimates by maximum likelihood are given by the diagonals of the matrix $-H^{-1}$ where $H$ represents an n x n matrix of second derivatives; n being the number of parameters in the model. The typical element of the matrix $H$, $H_{ab}$, is defined as

$$H_{ab} = \partial^2 L^*(\beta')/(\partial \beta_a.\partial \beta_b) \tag{3.61}$$

where $\beta'$ represents the vector of parameters estimated in the model and $L^*(\beta')$ represents the logarithm of the likelihood function that is maximised in order to obtain the values in

$\beta$'. That is,

$$L^{\cdot}(\beta') = \Sigma_i\Sigma_j T_{ij}\ln p_{ij}(\beta') \qquad (3.62)$$

where $T_{ij}$ represents the observed number of interactions between i and j and $p_{ij}(\beta')$ represents the model prediction of the probability of interaction between i and j. In the case of interaction models with only one parameter estimated by maximum likelihood, the computation of the variance of the estimate simplifies to:

$$\sigma^2(\beta') = - \partial L^{\cdot}(\beta')/\partial \beta^{2 \ -1} \qquad (3.63)$$

More details on the variance of maximum likelihood parameters are given by Mood and Graybill (1963) and by Cox (1970). Applications to spatial interaction modelling are provided by Giles and Hampton (1981) and computational notes are provided in Williams and Fotheringham (1984).

Because the MLEs are asymptotically normally distributed, their variances can be used to examine the significance of individual parameters in the usual manner. With maximum likelihood estimation, however, an alternative significance testing procedure exists through the construction of log-likelihoods. The statistic, $\tau$, defined as

$$\tau = 2.T.[L^{\cdot}(\beta') - L^{\cdot}(\beta';\beta_h = 0)] \qquad (3.64)$$

where T is the total volume of interaction is asymptotically chi-square distributed with 1 degree of freedom. The notation $L^{\cdot}(\beta';\beta_h = 0)$ represents the log-likelihood when the parameter whose significance is being examined is set to zero. A value of $\tau$ significantly different from zero allows one to reject the hypothesis that $\beta_h = 0$. Further details on the use of relative likelihood statistics in significance testing can be found in Hathaway (1975), Horowitz (1981) and Stopher and Meyburg (1979).

## 3.6  An Empirical Comparison of OLS and ML Parameter Estimates

Tobler (1983) reports data on 1970-1980 interregional migration between the nine major census divisions of the United States. These data, along with relevant distances and populations, are reported in Tables 3.1 and 3.2. Each of the four interaction models defined in Chapter 2 is calibrated by both OLS and MLE using these data. In all cases the intrazonal flows are ignored since the emphasis of the model calibration is on explaining interregional migration. Consequently, the doubly constrained model is calibrated by regression using equation (3.25) rather than (3.24). In this example, population is used as the sole measure of destination attractiveness and the effect of spatial separation is represented as a negative power function of distance. Thus, a maximum of three parameters can be estimated: an origin propulsiveness parameter, $\mu$; a destination attractiveness parameter, $\alpha$; and a distance-decay parameter, $\beta$. The estimates for each model are reported in Table 3.3.

It is clear that the OLS and MLE parameter estimates are not identical for any of the four models. This is due to different criteria involved in the OLS and MLE procedures. In OLS the criterion is to minimise the sum of squared differences between the predicted and actual independent variables; in MLE the criterion is to meet a non-linear constraint. The parameter estimates are similar, however, because when the ML constraints are met, it is likely that the predicted and observed interactions  will be similar and that the sum of squared differences between them will be near the minimum.

There is a noticeable trend in the estimated distance-decay parameters as constraints are added to the interaction model. This can be explained in terms of the ability of each model to replicate the set of migration flows. This ability is measured in Table 3.3 by the value of the goodness-of-fit statistic, SRMSE, defined in the subsequent section. The statistic has a value of zero when the observed flows are replicated perfectly and increasing values are indicative of increasingly poor model accuracy. Consequently, from Table 3.3 it is clear that adding constraints to the interaction model formulation increases accuracy, as would be expected.

**Table 3.1  Interregional Migration between Census Regions[*]**

|   | 1 | 2 | 3 | 4 | 5 | 6 | 7 | 8 | 9 |
|---|---|---|---|---|---|---|---|---|---|
| 1 | 0 | 180048 | 79223 | 26887 | 198144 | 17995 | 35563 | 30528 | 110792 |
| 2 | 283049 | 0 | 300345 | 67280 | 718673 | 55094 | 93434 | 87987 | 268458 |
| 3 | 87267 | 237229 | 0 | 281791 | 551483 | 230788 | 178517 | 172711 | 394481 |
| 4 | 29877 | 60681 | 286580 | 0 | 143860 | 49892 | 185618 | 181868 | 274629 |
| 5 | 130830 | 382565 | 346407 | 92308 | 0 | 252189 | 192223 | 89389 | 279739 |
| 6 | 21434 | 53772 | 287340 | 49828 | 316650 | 0 | 141679 | 27409 | 87938 |
| 7 | 30287 | 64645 | 161645 | 144980 | 199466 | 121366 | 0 | 134229 | 289880 |
| 8 | 21450 | 43749 | 97808 | 113683 | 89806 | 25574 | 158006 | 0 | 437255 |
| 9 | 72114 | 133122 | 229764 | 165405 | 266305 | 66324 | 252039 | 342948 | 0 |

[*] The census regions are: 1-New England, 2-Mid Atlantic, 3-East North-Central, 4-West North-Central, 5-South Atlantic, 6-East South-Central, 7-West South-Central, 8-Mountain, and 9-Pacific.

**Table 3.2  Populations and Distances in Miles between Region Centroids**

| Region | Population | 1 | 2 | 3 | 4 | 5 | 6 | 7 | 8 | 9 |
|---|---|---|---|---|---|---|---|---|---|---|---|
| 1 | 11848000 | - | | | | | | | | |
| 2 | 37056000 | 219 | - | | | | | | | |
| 3 | 40266000 | 1009 | 831 | - | | | | | | |
| 4 | 16327000 | 1514 | 1336 | 505 | - | | | | | |
| 5 | 29920000 | 974 | 755 | 1019 | 1370 | - | | | | |
| 6 | 13096000 | 1268 | 1049 | 662 | 888 | 482 | - | | | |
| 7 | 19025000 | 1795 | 1576 | 933 | 654 | 1144 | 662 | - | | |
| 8 | 8289000 | 2420 | 2242 | 1451 | 946 | 2278 | 1795 | 1278 | - | |
| 9 | 25476000 | 3174 | 2996 | 2205 | 1700 | 2862 | 2380 | 1779 | 754 | - |

The variation in the estimated distance-decay parameter between the models is thus probably related to the variation in the accuracy of the models. Because the doubly constrained model is most accurate it could be assumed that the estimate derived from this model is the most accurate representation of the true relationship between interaction and distance. The difference between the distance-decay parameter estimates of the two singly-constrained models can also be explained in terms of the accuracy of the two models. The attraction-constrained model is more accurate in replicating migration flows than the production-constrained model and hence the parameter estimate of the former is more similar to that of the doubly constrained model. A possible reason for the superior performance of the attraction-constrained model is that the population size of a region is probably a more accurate measure of propulsiveness than of attractiveness. Many other variables such as climate, economic factors and quality-of-life determine the attractiveness of a region for migration. Such variables do not seem to be as important in determining the number of people leaving particular regions.

Table 3.3  **OLS and ML Parameter Estimates for Four Interaction Models**

| Model | OLS Results | | | | MLE Results | | | |
|---|---|---|---|---|---|---|---|---|
| | $\mu$ | $\alpha$ | $\beta$ | SRMSE | $\mu$ | $\alpha$ | $\beta$ | SRMSE |
| Uncon-strained | .828 | .742 | .452 | .596 | .692 | .635 | .367 | .583 |
| Production-constrained | * | .639 | .572 | .563 | * | .658 | .494 | .560 |
| Attraction-constrained | .703 | * | .709 | .343 | .737 | * | .718 | .342 |
| Doubly-constrained | * | * | .994 | .245 | * | * | .905 | .234 |

Through the use of equation (3.15), it is possible to employ the data presented in Tables 3.1 and 3.2 to derive estimates of the relative attractiveness of each of the nine regions for migrants in the United States. These results are reported in Table 3.4 where each value is an estimate of attractiveness relative to the East North-Central region which was set to 1.0 (see discussion above). Clearly, by far the most attractive regions are the Pacific region and the South Atlantic region while the least attractive regions are the North-East and the West North-Central.

Table 3.4 **Estimates of Relative Attractiveness**

| Number | Region | Estimated Relative Attractiveness |
|--------|--------|-----------------------------------|
| 1 | North-East | 0.39 |
| 2 | Mid-Atlantic | 0.82 |
| 3 | East North-Central | 1.00 |
| 4 | West North-Central | 0.60 |
| 5 | South Atlantic | 2.25 |
| 6 | East South-Central | 0.42 |
| 7 | West South-Central | 1.03 |
| 8 | Mountain | 0.98 |
| 9 | Pacific | 3.90 |

Other empirical comparisons of OLS and MLE parameter estimates in interaction modelling can be found in Stetzer (1976), van Est and van Setten (1979), Openshaw (1979) and Fotheringham and Williams (1983).

### 3.7 Goodness-of-Fit Statistics and Spatial Interaction Models

An important component of spatial interaction modelling is the assessment of a model's ability to replicate a known set of flows. Accurate replication supports the theoretical propositions on which the model is based: that is, it supports one particular model form over others. It also lends confidence in the accuracy of parameter estimates and in the ability of a model to predict flows in systems other than that in which the model was calibrated. Many goodness-of-fit statistics have been employed in spatial interaction modelling to assess a model's ability to replicate a data set and reviews are presented by Knudsen and Fotheringham (1986) and by Fotheringham and Knudsen (1987). All such statistics involve a quantitative description of some aspect of the difference between $T'$, the matrix of predicted flows, and $T$, the matrix of observed flows, but there has been little consistency in the use of a particular statistic(s) which hinders comparison of model performance across studies (see Hathaway, 1975; Thomas, 1977; Southworth, 1983; and Fotheringham and Williams, 1983). It is also unfortunate that the use of different goodness-of-fit statistics can lead to different conclusions being reached regarding model performance. For example, Willmott (1984) employs three goodness-of-fit measures to evaluate model performance (although not in a spatial interaction context), and each statistic indicates a different model as the most accurate.

In the light of the research on goodness-of-fit by Knudsen and Fotheringham (1986) and Fotheringham and Knudsen (1987), a reasonable strategy to employ in evaluating spatial interaction models would appear to be to employ a combination of two of the following three statistics: $R^2$, Information Gain and SRMSE, the Standardised Root Mean Square Error. The statistic $R^2$ and its properties are well-known, and its significance can be examined through a t-test. Information Gain is slightly less well-known and is calculated as:

$$I = \sum_i \sum_j T_{ij} \ln(T_{ij}/T_{ij}') \tag{3.65}$$

It has a lower value of zero corresponding to a perfect set of predictions and upper limit of infinity. Its significance can be found through its relationship to the minimum discrimination information (MDI) statistic,

$$MDI = 2T.I \tag{3.66}$$

which is asymptotically chi-square distributed (Bishop et al., 1975; Phillips, 1981) with mn-k degrees of freedom, where mn represents the number of origin-destination pairs and k represents the number of parameters in the model. SRMSE is calculated as

$$SRMSE = (1/T)[\sum_i \sum_j (T_{ij} - T_{ij}')^2/n] \tag{3.67}$$

It has a lower limit of zero, indicating a completely accurate set of predictions, and an upper limit that, although variable and dependent on the distribution of the observed flows, is usually 1.0. Values of SRMSE greater than 1.0 only occur when the mean error in predicting a set of flows is greater than the mean flow. Unfortunately, SRMSE does not have a known theoretical distribution so that significance testing can only be carried out through elaborate experimental procedures (see Fotheringham and Knudsen, 1987). However, its value has a strongly linear relationship with our general concept of error which makes it useful as a comparative measure of model performance.

On theoretical grounds $R^2$ is more applicable to assessing the goodness-of-fit of linear models and so can be employed to examine the performance of models made linear by the transformations described above in Section 3.2. Information Gain is most applicable to models calibrated by maximum-likelihood and SRMSE can be applied to any models calibrated by any means. Thus, a useful combination of goodness-of-fit statistics to report for linear interaction models is $R^2$ and SRMSE; while a useful combination for models calibrated by MLE is Information Gain and SRMSE.

Examples of the use of the above statistics in spatial interaction modelling include Fotheringham's (1983a) use of $R^2$ to assess differences in linear interaction model specification; Lewis's (1975) use of $R^2$ to assess differences in model performance between gravity and Heckscher-Ohlin models of interregional trade; the use of SRMSE by Pitfield (1978) to discriminate between a linear programming model and a doubly-constrained interaction model of freight movements in Britain; and the use of Information Gain by Thomas (1977) in a study of journeys-to-work on Merseyside. Fotheringham and Williams (1983) report all three of the above statistics to compare the calibration results of a production-constrained model by OLS and MLE in four different data sets. Reassuringly, the three statistics had the same order of relative magnitude in all four data sets.

Many goodness-of-fit statistics, other than those reported above, can be employed in spatial interaction modelling and the reader is referred to Fotheringham and Knudsen (1987) for a survey of such statistics. However, in the interests of comparability and consistency, we recommend that a combination of the above three statistics always be reported.

### 3.8   The Use of SIMODEL to Calibrate Spatial Interaction Models

While the calibration of interaction models by least squares regression can easily be undertaken with one of a multitude of computer software packages, maximum-likelihood calibration packages are less frequently encountered. As a consequence, an overview is now presented of a package, SIMODEL, that has been written for the exclusive purpose of calibrating spatial interaction models by MLE.

SIMODEL (Williams and Fotheringham, 1984) is a FORTRAN program which has been extensively tested in classroom and research situations and has proven to be a robust and efficient algorithm. To give an idea of the execution time of SIMODEL, the calibration of a two-parameter model with a 25 x 25 interaction matrix takes 1.76 seconds of CPU time on a CDC Cyber 170/855. SIMODEL is now implemented at research sites throughout the world.

Table 3.5  **Options Available for Calibration in SIMODEL**

| OPTION | MODEL DESCRIPTION | FORMAT | NO. OF PARAMETERS |
|---|---|---|---|
| 1 | Production-constrained | $T_{ij} = A_i O_i D_j f(d_{ij})$ | 1 |
| 2 | Production-constrained | $T_{ij} = A_i O_i W_j f(d_{ij})$ | 2 |
| 3 | Production-constrained | $T_{ij} = A_i O_i D_j f(d_{ij})c_j^\delta$ | 2 |
| 4 | Production-constrained | $T_{ij} = A_i O_i W_j f(d_{ij})c_j^\delta$ | 3 |
| 5 | Destination-constrained | $T_{ij} = B_j O_i D_j (fd_{ij})$ | 1 |
| 6 | Destination-constrained | $T_{ij} = B_j v_i^\mu D_j f(d_{ij})$ | 2 |
| 7 | Poisson unconstrained | $T_{ij} = v_i^\mu w_j^\alpha f(d_{ij})$ | 3 |
| 8 | Poisson unconstrained | $T_{ij} = v_i^\mu w_j^\alpha f(d_{ij})c_j^\delta$ | 4 |
| 9 | Doubly constrained | $T_{ij} = A_i B_j O_i D_j f(d_{ij})$ | 1 |

**Notes on Table 3.5**

In the above notation, $T_{ij}$ represents the interaction between origin i and destination j;  $O_i$ represents the known total outflow from i; $D_j$ represents the known total inflow into j; $v_i$ represents origin propulsiveness; $w_j$ represents destination attractiveness; $d_{ij}$ represents the separation between i and j; and $c_j$ represents the competition destination j faces from all other destinations and is usually measured by the accessibility of j to all other destinations (see Chapter 4 for more details). $A_i$ and $B_j$ are balancing factors. The parameters  $\mu$,  $\alpha$, and  $\delta$  are estimated in the model calibration. A distance-decay parameter, ß, is also calibrated in the separation function, $f(d_{ij})$.

The data input to SIMODEL has a very simple structure consisting of one logical record describing the model option chosen and the structure of the data to be read followed by the data themselves. The logical record consists of the following eight elements:

1. Model option to be calibrated (see Table 3.5)
2. Distance/Cost function to be used (power or exponential)
3. Number of origin-destination pairs for which data are available
4. Number of origins
5. Number of destinations
6. Minimum distance declaration. Only interactions over distances greater than this value are used in the model calibration
7. Maximum iteration limit. A value of 100 is suggested
8. Probability data declaration. A value of one is entered if the interaction data are in the form of probabilities; 0 otherwise.

After reading the above record, SIMODEL then reads interaction and distance data which are entered on one logical record per origin-destination pair. Following this, data on origins and destinations are read in as blocks. As an example, the following SIMODEL program would calibrate a Poisson unconstrained model with a power distance function for a 2 x 2 interaction model using all flows over a distance greater than zero units. The interaction and distance data are read in first, followed by an origin propulsiveness measure and two destination variables. Four parameters would then be estimated in the model calibration.

```
6POWER      4     2     2   0.0   100     0
1999.0        23.0     1    1
 423.0       124.1     1    2
 627.0        99.2     2    1
3241.0        11.2     2    2
110682.0
342891.0
221264.0
684682.0
  834.89
  253.71
```

In terms of output, SIMODEL produces a series of calibration results plus a range of diagnostic information. These are demonstrated in Table 3.6 for a production-constrained model calibrated with data on airline passenger flows from Atlanta, Georgia to 25 other cities in the United States. The model has two parameters to be estimated: a destination attractiveness parameter and a distance-decay parameter. These are reported with confirmation that their respective constraint equations are met. This is followed by some general information on the interaction matrix and then a series of goodness-of-fit measures. The standard errors of the parameter estimates are then reported plus a series of log-likelihood values that can also be used to assess the significance of each parameter estimate. Following that, the balancing factors (only one in this case) are printed plus the observed and predicted flows and absolute and percentage errors. Further details on SIMODEL output can be found in Williams and Fotheringham, 1984.

**Table 3.6 An Example of SIMODEL Output for a Production-Constrained Interaction Model**

```
SIMODEL VERSION 1.0 UNDER NOS 2.2 - RELEASED JULY 1984


PRODUCTION CONSTRAINED GRAVITY MODEL RESULTS WITH OPTION  2
FOR    1 ORIGIN(S) AND    25 DESTINATION(S) CONSIDERING
ONLY INTERACTIONS AT DISTANCES OR COSTS GREATER THAN   160.0


THE OBSERVED MEAN TRIP BENEFIT  =     14.9831
THE PREDICTED MEAN TRIP BENEFIT =     14.9831

MAXIMUM LIKELIHOOD MASS PARAMETER            =    .7818262


THE OBSERVED MEAN TRIP LENGTH    =     6.5003
THE PREDICTED MEAN TRIP LENGTH  =     6.5003

MAXIMUM LIKELIHOOD DISTANCE PARAMETER     =   -.7365098


AFTER    10 ITERATIONS OF THE CALIBRATION ROUTINE
WITH A POWER DISTANCE OR COST FUNCTION


THE NUMBER OF ORIGIN-DESTINATION PAIRS CONSIDERED =    24

THE TOTAL INTERACTIONS OBSERVED  =     242873.0
THE TOTAL INTERACTIONS PREDICTED =     242873.0

THE ASYMMETRY INDEX FOR THIS INTERACTION DATA =      0


REGRESSING THE OBSERVED INTERACTIONS ON THE PREDICTED
INTERACTIONS YIELDS AN R SQUARED VALUE OF    .605

TIJ (OBS) =    -1235.6 +  1.122 TIJ (PRED)
                    (    .195)

T STATISTIC FOR REGRESSION (SIG. DIFF. PARA. FROM 1) =    .6262


PERCENTAGE DEVIATION OF OBSERVED INTS. FROM THE MEAN (  10119.7) = 57.965

PERCENTAGE DEVIATION OF PREDICTED INTS. FROM THE OBSERVED INTS.  = 42.189

PERCENTAGE REDUCTION IN DEVIATION = 27.216


AYENI S INFORMATION STATISTIC (PSI) =   .617062E-01

MINIMUM DISCRIMINANT INFORMATION STATISTIC =   .299735E+05

THE STANDARDIZED ROOT MEAN SQUARE ERROR STAT.=   .5787
```

```
THE MAXIMUM ENTROPY FOR      24 CASES        =    3.1781
THE ENTROPY OF THE PREDICTED INTERACTIONS    =    3.0063
THE ENTROPY OF THE OBSERVED INTERACTIONS     =    2.8736

MAXIMUM ENTROPY - ENTROPY OF PREDICTED INTS. =    .1718

ENTROPY OF PRED. INTS.- ENTROPY OF OBS. INTS.=    .1327

ENTROPY RATIO STATISTIC =   .5641


VARIANCE OF THE ENTROPY OF PREDICTED INTERACTIONS =   .165272E-05

VARIANCE OF THE ENTROPY OF OBSERVED INTERACTIONS  =   .255475E-05

T STATISTIC FOR THE ABSOLUTE ENTROPY DIFFERENCE =    64.6963

THE INFORMATION GAIN STATISTIC =    .13270573


AVERAGE DISTANCE TRAVELED IN SYSTEM =    736.5283
AVERAGE ORIGIN-DESTIN. SEPARATION   =    851.2500


STANDARD ERROR OF MLE DISTANCE PARAMETER    =   .527456E-02

STANDARD ERROR OF MLE MASS PARAMETER        =   .276766E-02


THE LOG-LIKELIHOOD VALUE OF THE FITTED MODEL WITH ALL
PARAMETERS    =   -.300629E+01


THE LOG-LIKELIHOOD VALUE OF THE FITTED MODEL WITHOUT
THE PARAMETER     .7818262 =   -.317735E+01


THE RELATIVE LIKELIHOOD (LAMBDA) STATISTIC FOR THE
PARAMETER          .7818262 =    .830900E+05


THE LOG-LIKELIHOOD VALUE OF THE FITTED MODEL WITHOUT
THE PARAMETER    -.7365098 =   -.305742E+01


THE RELATIVE LIKELIHOOD (LAMBDA) STATISTIC FOR THE
PARAMETER         -.7365098 =    .248337E+05


THE LOG-LIKELIHOOD VALUE OF THE FITTED MODEL WITHOUT
ALL THE MODEL PARAMETERS   =   -.317805E+01


THE RHO-SQUARED STATISTIC FOR THE MODEL    =   .540466E-01


THE ADJUSTED RHO-SQUARED STATISTIC         =   -.319492E-01


THE LOG-LIKELIHOOD VALUE OF THE MEAN MODEL= -.317805E+01
```

| I | A(I) BALANCING FACTOR | O(I) OBSERVED OUTFLOW |
|---|---|---|
| 1 | .485498E-04 | 242873.0 |

| J | OBSERVED INFLOWS | PREDICTED INFLOWS |
|---|---|---|
| 1 | 0 | 0 |
| 2 | 6469.0 | 9476.4 |
| 3 | 7629.0 | 9635.4 |
| 4 | 20036.0 | 23858.4 |
| 5 | 4690.0 | 9312.6 |
| 6 | 6194.0 | 9662.2 |
| 7 | 11688.0 | 9078.8 |
| 8 | 2243.0 | 3675.5 |
| 9 | 8857.0 | 16637.3 |
| 10 | 7248.0 | 8052.7 |
| 11 | 3559.0 | 5729.4 |
| 12 | 9221.0 | 10110.4 |
| 13 | 10099.0 | 6983.1 |
| 14 | 22866.0 | 6305.0 |
| 15 | 3388.0 | 6515.2 |
| 16 | 9986.0 | 6949.6 |
| 17 | 46618.0 | 30924.7 |
| 18 | 11639.0 | 16384.9 |
| 19 | 1380.0 | 2480.7 |
| 20 | 5261.0 | 11373.4 |
| 21 | 5985.0 | 12127.7 |
| 22 | 6731.0 | 4621.1 |
| 23 | 2704.0 | 2652.7 |
| 24 | 12250.0 | 7364.3 |
| 25 | 16132.0 | 12961.5 |

| ORIGIN | DESTINATION | OBSERVED FLOW | PREDICTED FLOW | ABSOLUTE ERROR | PERCENT ERROR |
|---|---|---|---|---|---|
| 1 | 2 | 6469.0 | 9476.4 | 3007.4 | 46.49 |
| 1 | 3 | 7629.0 | 9635.4 | 2006.4 | 26.30 |
| 1 | 4 | 20036.0 | 23858.4 | 3822.4 | 19.08 |
| 1 | 5 | 4690.0 | 9312.6 | 4622.6 | 98.56 |
| 1 | 6 | 6194.0 | 9662.2 | 3468.2 | 55.99 |
| 1 | 7 | 11688.0 | 9078.8 | -2609.2 | -22.32 |
| 1 | 8 | 2243.0 | 3675.5 | 1432.5 | 63.87 |
| 1 | 9 | 8857.0 | 16637.3 | 7780.3 | 87.84 |
| 1 | 10 | 7248.0 | 8052.7 | 804.7 | 11.10 |
| 1 | 11 | 3559.0 | 5729.4 | 2170.4 | 60.98 |
| 1 | 12 | 9221.0 | 10110.4 | 889.4 | 9.65 |
| 1 | 13 | 10099.0 | 6983.1 | -3115.9 | -30.85 |
| 1 | 14 | 22866.0 | 6305.0 | -16561.0 | -72.43 |
| 1 | 15 | 3388.0 | 6515.2 | 3127.2 | 92.30 |
| 1 | 16 | 9986.0 | 6949.6 | -3036.4 | -30.41 |
| 1 | 17 | 46618.0 | 30924.7 | -15693.3 | -33.66 |
| 1 | 18 | 11639.0 | 16384.9 | 4745.9 | 40.78 |
| 1 | 19 | 1380.0 | 2480.7 | 1100.7 | 79.76 |
| 1 | 20 | 5261.0 | 11373.4 | 6112.4 | 116.18 |
| 1 | 21 | 5985.0 | 12127.7 | 6142.7 | 102.64 |
| 1 | 22 | 6731.0 | 4621.1 | -2109.9 | -31.35 |
| 1 | 23 | 2704.0 | 2652.7 | -51.3 | -1.90 |
| 1 | 24 | 12250.0 | 7364.3 | -4885.7 | -39.88 |
| 1 | 25 | 16132.0 | 12961.5 | -3170.5 | -19.65 |

# CHAPTER 4

## SPATIAL INTERACTION AND SPATIAL CHOICE

### 4.1 Introduction

Most spatial interaction results from some sort of spatial choice. Whether it be an individual consumer selecting a store, an individual selecting a residence (and, hence determining commuting patterns), or a migrant selecting a city in which to live, in each case the resulting interaction depends upon a spatial choice. Consequently, there is a strong relationship between modelling spatial interaction and modelling spatial choice which is explored in this chapter for three main reasons:

(i) There exists a large volume of research on discrete choice modelling, both spatial and aspatial, and it is useful to establish how this relates to the even larger volume of research on spatial interaction modelling;

(ii) The discrete choice framework provides an alternative theoretical justification for the form of the two singly constrained spatial interaction models. It was shown in Chapter 1 that the production-constrained interaction model represents the choices of destinations (shops, cities etc.) while the attraction-constrained model represents the choices of origins (residences). Although the theoretical development of these models has already been presented in Chapter 2 in terms of information theory, the discrete choice framework provides a more behavioural, and possibly more acceptable, framework for understanding the rationale of these models.

(iii) Understanding the behavioural foundation of spatial choice models allows the identification of possible shortcomings in this foundation and can lead to improvements in model formulation. A demonstration of this is provided in terms of modelling hierarchical destination choice which is used to suggest a more general framework in which to model spatial choice and hence spatial interaction.

Essentially, the basic difference between spatial interaction and spatial choice models is one of usage: by consensus, the term spatial interaction is applied to aggregate flows whereas the term spatial choice is applied to an individual selection of a location. Clearly, aggregate flows are the result of a collection of individual decisions so the two are inextricably linked: the variables that explain the spatial choice of an individual tend to be very similar to the variables that explain the spatial choices of a large number of individuals. However, by focussing on the level of the individual, spatial choice is clearly more behavioural and indeed this chapter contains much which can be classified as "geo-psychological". This term is employed to describe a type of analysis linking spatial choice to spatial awareness and spatial information processing. In turn, given the broader aims of this book, the relationship between spatial choice and spatial interaction is developed.

### 4.2 The General Spatial Choice Problem

While the examples of spatial choices given above differ greatly in certain aspects (compare, for example, the choice of a store for grocery shopping with the choice of an urban area in which to live), they all share the following common features:

(i) The choice process is discrete in that an alternative is either selected or not selected, no intermediate possibility exists;

(ii) The alternatives have fixed spatial locations so that they form a spatial pattern with alternatives facing various degrees of competition from one another;

(iii) Each alternative is a fixed distance from the individual making a choice and usually the set of distances between the individual and the alternatives exhibit strong internal variation;

(iv) Each alternative can be defined in terms of a set of attributes. These attributes can be divided into three types: **site attributes**, which describe the features of alternatives that are independent of location; **situation attributes,** which describe the features of alternatives that are derived from their relative location with respect to other alternatives; and **separation attributes** that represent the spatial separation between the individual and the alternative.

The general spatial choice problem can thus be described as in Figure 4.1 and stated thus: Given that an individual i is faced with making a choice from a set N consisting of n spatial alternatives each labelled j, how is a particular alternative, j', selected?

Figure 4.1  **The Spatial Choice Problem**

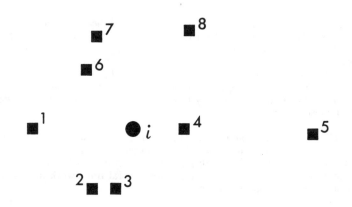

### 4.3  Spatial and Aspatial Choice

Spatial choice is clearly a discrete phenomenon: that is, the outcome of the choice process is such that an alternative has either been selected or it has not - there is no intermediate possibility. Consequently, it is only natural that much research into spatial choice has relied upon the early discrete choice modelling advances made in areas such as brand choice in marketing and mode choice in transportation (Hensher and Johnson, 1981; Ben-Akiva and Lerman, 1985; Train, 1986) which are aspatial. There is no spatial component, for example, involved in a consumer's choice of coffee brand or in the selection of bus over train. Such choices do not conform to the second, third and fourth attributes of the general spatial choice problem described above.

While there is nothing inherently wrong with the transference of ideas between disciplines, and in fact it is usually beneficial, sufficient difference exists between spatial and aspatial choice to warrant concern in this particular case. Fotheringham (1988b) has outlined several differences between spatial and aspatial choice that have potential implications for the way in which choice is modelled.

In spatial choice situations, the number of alternatives is often  very large compared to aspatial choice.  Consider, for example, the   difference between a migrant selecting an urban area where there may  be thousands of alternatives, and a consumer selecting a mode of  transportation where there may only be two or three alternatives.

In certain choice situations it is clear that individuals perceive  alternatives in groups or clusters, a point expanded subsequently.   It is usually easier to identify such clusters in aspatial choice,  where examples  include brands of diet soft drinks versus regular  soft drinks, or public versus private transit,  than it is in  spatial choice.  This difference arises because, whereas in aspatial  choice the discriminating factor between clusters is discrete (transit is either public or private), the discriminating factor  between clusters of spatial alternatives is usually space itself  which is continuous.  If one were to impose discrete boundaries  around clusters of destinations, this would likely result in a  situation where an alternative near the border of one cluster and an   alternative near the border of a neighbouring cluster are not  considered substitutes for each other, yet are located in close proximity.  The imposition of discrete boundaries in such an  instance is usually subjective and likely to result in errors in the  subsequent modelling procedure.  Boundaries of spatial clusters are  more likely to be  "fuzzy" (Zadeh, 1965; Gale, 1972; Pipkin, 1978)  than discrete which causes problems in the applications of certain  discrete choice models such as the nested logit that need an <u>a priori</u> definition of clusters.  The successful application of such  models in an aspatial context thus offers no guarantee of similar  success in a spatial context.

In an aspatial context, within any one identifiable cluster of  alternatives such as diet soft drinks there is an implicit  assumption of randomness in the arrangement of alternatives. That is, the alternatives are assumed to be equal substitutes for one  another.  As an example, if a store is out of one desired brand of  diet soft drink, the consumer will probably substitute another brand  of diet soft drink which, from the modeller's perspective, will  appear entirely arbitrary.  In a spatial choice cluster, however,  there is an ordering of alternatives caused by the different spatial  relationships that exist for each alternative. One cannot  arbitrarily "move" alternatives around without changing the context  of the choice process.  Clearly, the more two alternatives are  separated by space, the less likely they are to be substitutes for  each other, which again reflects the continuous nature of most spatial relationships.  Formally, if $S_{j'j}$ is defined as the degree  of substitution between two alternatives, j' and j, the following  relationship is generally implicitly if not explicitly assumed in   aspatial choice:

$$S_{j'j} = \begin{cases} 1 \text{ if j' and j are in the same cluster} \\ 0 \text{ otherwise} \end{cases} \tag{4.1}$$

However, in spatial choice

$$S_{j'j} = f(d_{j'j}) \tag{4.2}$$

where f(  ) represents a continuous decreasing function and $d_{j'j}$  represents the spatial separation between j' and j.  That is, as the  distance between two alternatives increases, the degree to which an  individual may substitute one for the other, decreases.  Thus, if a consumer finds that one store does not have a desired product and  there are no suitable substitutes, he/she is likely to select  another store in close proximity to the initially-selected store.  The choice of a substitute store will not appear arbitrary.  It can  be seen from equations (4.1) and (4.2) that while there is  transitivity between the substitution of aspatial choices, there is  no guarantee of a similar transitivity in spatial choice.  Consider,  for example, three brands of coffee, A, B and C.  Suppose that brand  A and B are substitutes for one another and that brands B and C are  also substitutable.  It then follows that brands

A and C are substitutes, or, at least, there is a high probability that they are. This same transitivity is much less likely to occur in a spatial setting. Consider three destinations, X, Y and Z, located in a cluster and that Y lies midway between X and Z. If X and Y and Y and Z are substitutes, then because of their greater spatial separation, there is no guarantee that X and Z are substitutes.

Finally, in aspatial choice, the set of alternatives forming each cluster is usually assumed to be constant across individuals. For example, individuals at different locations are likely to perceive membership of the clusters of regular and decaffeinated coffee in the same way. This assumption of cluster consistency is very unlikely to occur in spatial choice and evidence from the literature on mental maps (Gould and White, 1974) strongly suggests that individuals in different locations have different perceptions of space and hence of spatial clusters.

These differences between spatial and aspatial choice are shown below to have important implications for the application of certain spatial choice models. In particular, it appears that it is not always possible to transfer the developments in discrete choice modelling from an aspatial context to a spatial one.

### 4.4   Attributes of Spatial Alternatives

Implicit in the solution to the general spatial choice problem outlined above is the assumption that each alternative can be defined in terms of a set of attributes $\{X: x_k, k=1,...,r\}$ where the index k denotes a particular attribute. The dominant theme in much spatial choice modelling and spatial interaction modelling to date has been the investigation of the composition of the set X: that is, the definition of the relevant attributes of alternatives that affect spatial choice. There is a substantial literature, for example, on this topic in store choice (inter alia, Downs, 1970; Hansen and Deutscher, 1977; Recker and Kostyniuk, 1978; Hubbard, 1978; Schuler, 1979; McCarthy, 1980; and Gautschi, 1981), migration (inter alia Greenwood and Sweetland, 1972; Dunlevy and Gemery, 1977; Clark and Ballard, 1980; Hervitz, 1983; and Op't Veld et al., 1984) and residential choice (inter alia Peterson, 1967; Clark and Cadwallader, 1973; Johnston, 1973; Cadwallader, 1979; and Preston and Taylor, 1981). While a detailed discussion of the rationale for the inclusion of particular attributes in certain spatial choice situations is left to subsequent chapters, a general comment relating to these attributes is worth noting. What becomes clear in reading the many investigations of the attributes affecting spatial choice is the great diversity of attributes that appear to be relevant across different choice situations. Indeed, even within the same context, the set of relevant attributes may vary across space or across socio-economic groups. For example, the set of destination attributes affecting migration in the United States is unlikely to be the same as the set of attributes affecting migration in the Netherlands. Climatic variations, for instance, important in a U.S. context, are virtually non-existent in the Dutch context. Similarly, the determinants of store choice by the elderly may not be the same as those by the young. Thus, the investigation of attribute sets, while of obvious importance in particular choice contexts, can play only a limited role in the development of general theories of spatial choice. Advances in understanding the bases for spatial choice must therefore come from some other direction: namely from understanding the way in which individuals process the information on attributes in order to reach a spatial choice.

### 4.5   Information Processing and Spatial Choice

Two uncertainties exist in modelling spatial choice: one concerns the attributes of destinations that make them attractive or unattractive to an individual and, as discussed above, while this has been the focus of the majority of research in spatial choice modelling, it has limited general application; the other concerns the way in which individuals process information to arrive at a choice given that the relevant destination attributes have been identified. This latter uncertainty is now considered in some detail because the way in which individuals process information to arrive at a choice determines the type of model that should be employed to understand and predict such choices.

Various information - processing strategies are listed in Table 4.1. The simplest is a purely random one whereby an individual selects an alternative in a completely arbitrary manner. Following such a strategy would obviously lead to unsatisfactory choices being made in most instances. It therefore would seem to be a strategy that is not employed very frequently - particularly not in spatial choice where there is a great deal of empirical evidence to suggest that individuals are constrained by distance effects and exhibit a strong tendency to select alternatives in close proximity. If individuals were to make spatial choices randomly, such choices could be modelled with a Bernoulli process (Burnett 1974).

Table 4.1 **Information-Processing Strategies**

| Simple | Non-Compensatory | Compensatory | Hybrid |
|--------|------------------|--------------|--------|
| Random | Conjunctive | Optimising | Conjunctive/Optimising |
| | Disjunctive | Satisficing | Disjunctive/Optimising |
| | Lexicographic | Hierarchical | |
| | Sequential Elimination | | |

The second group of processing strategies, termed non-compensatory, assume that alternatives can be differentiated on the basis of a subset of attributes and that adverse scores on one attribute cannot be compensated by favourable scores on another. In the conjunctive case an individual sets minimum thresholds on all attributes and if each attribute of an alternative does not exceed the threshold (for that particular attribute), the alternative is eliminated from consideration. These thresholds can be raised until only one alternative remains. In the disjunctive case an individual sets high thresholds for each attribute and if any attribute of an alternative exceeds its particular threshold, that alternative is selected. Thus, in the conjunctive case, negative attributes are more important in determining selection; in the disjunctive case, positive attributes are more important.

In the lexicographic strategy, an individual orders the attributes in terms of importance and selects the alternative having the highest score on the most important attribute. An important set of spatial choice models, nearest-center models, is based on the assumption of a lexicographic information-processing strategy. In these models, exemplified by linear programming, location-allocation and central place theory, individuals are assumed to rank distance to the alternative as the most important attribute and then select the alternative with the optimum (in this case, minimum) value of this attribute. Such models are discussed further in Chapter 8.

In the fourth non-compensatory strategy, that of sequential elimination, minimum thresholds are established for each attribute; an attribute is then selected at random and alternatives are eliminated that fail to meet the threshold for that attribute. The process continues until one alternative remains. A special case of sequential elimination in which a probability function for selection is attached to each attribute is the foundation of the Elimination-by-Aspects model (Tversky 1972) and the Elimination-by-Tree   model (Tversky and Sattath 1979). Such models have been criticised (inter alia, Kahn et al. 1985, p.10) for being very cumbersome and of little practical use. Indeed, in modelling spatial choice in particular, the non-compensatory, information-processing strategies would seem to be unrealistic. There is a large amount of empirical evidence in the retailing literature, for example, which suggests that there is a trade-off between a small store in close proximity and a large, distant store.  Hubbard (1978) reports results that only 38 per cent of a consumer sample in Worcester, Mass, patronised the nearest store (Thompson, 1967), only 50 per cent did in Christchurch, New Zealand (Clark, 1968), and only 25 per cent did in Manchester, U.K. (Fingleton, 1975). Hubbard's general conclusion on the results of empirical tests of the nearest-center hypothesis was that "while the friction of distance factor is obviously important in explaining the travel behaviour of consumers, it appears equally obvious that its effects should be incorporated in a more realistic fashion than that provided by the nearest center postulate". (p.4)

The  trade-off  between  attribute  values  is  a  feature  of  the  compensatory information-processing strategies. In each procedure, attributes are   combined in some manner so that an overall score is assigned to each alternative; the score representing an individual's utility or benefit from choosing that alternative. In the optimising procedure, every alternative is given a score in this manner and the alternative having the highest score is selected. In this way, the procedure guarantees that within the constraints of the information available to the individual, he/she will make the optimal spatial choice in terms of maximising utility. In the other two compensatory strategies, the individual does not necessarily evaluate every alternative and assign it a score. In the satisficing procedure, alternatives are evaluated in a random order until an alternative that is deemed "satisfactory" in terms of its overall score is found. In the hierarchical strategy the alternatives are assumed to be perceived by the individuals as being grouped into clusters which are then evaluated. Only within a selected cluster are the individual alternatives then evaluated according to the optimising procedure. There is no guarantee that an optimal choice will be made from either the satisficing or hierarchical strategies but both have the advantage of reducing the number of evaluations that are necessary in order to make a spatial choice.

The hybrid processing strategies are fairly self-evident. The conjunctive/optimising strategy assumes a two-stage process whereby a conjunctive strategy is first employed to produce a subset of alternatives that exceed the thresholds placed on all attributes and then an optimising   procedure  is  employed  to  select  one  of  these  alternatives.  The disjunctive/optimising strategy assumes a similar process except that in the first process a pool of alternatives that exceed any of the thresholds placed on the attributes is selected.

Clearly the concept of information-processing is important to the modelling of spatial choice. Different processes should be modelled in different ways. For instance, as will be shown, the optimising procedure is the basis of the logit formulation (McFadden, 1974a; 1974b) and, hence, the singly constrained interaction models; the satisficing process leads to a hypogeometric model (Beavon and Hay, 1977); and a hierarchical process is the basis for both the nested logit formulation (McFadden, 1978; 1980) and the competing destinations model (Fotheringham, 1983a; 1986).

However, it is not the task of this chapter to provide a comprehensive review of spatial choice models since such a review can be found elsewhere (inter alia van Lierop and Nijkamp, 1982; Wrigley, 1985). Rather, the concentration is on exploring the relationship between information processing and spatial choice modelling more closely and for this purpose the behavioural assumptions embedded in three models, the logit, nested logit and competing destinations model, will be examined. Each of these models can be derived theoretically within a random utility maximisation framework and so a general discussion of this framework is first given in the context of spatial choice.

## 4.6 A Theoretical Justification for the Form of Singly Constrained Interaction Models in Terms of Random Utility Maximisation

Because the outcome of a spatial choice is discrete, an alternative being either selected or rejected, it is convenient to turn to the well-established tenets of discrete choice theory and, in particular, the random-utility maximisation framework, to investigate spatial choice. However, sufficient differences exist between the more traditional aspatial choice and spatial choice to warrant concern in the wholescale adoption of traditional discrete choice theory. Consequently, while the traditional theory is briefly reviewed, refinements are made to this theory to produce a more realistic overall framework for modelling spatial choice.

Define the level of utility or benefit accruing to individuals from choosing alternative j' as $U_{ij'}$ and define $U_{ij'}$ as

$$U_{ij'} = V_{ij'} + \mu_{ij'} \tag{4.3}$$

where $V_{ij'}$ is the observable component of an individual's utility and is a function of the attributes of a particular alternative, and $\mu_{ij'}$ is a random error term. Define $p_{ij'}$ as the probability that individual i selects alternative j'. It is assumed that the individual processes information according to the optimising principle and after evaluating all alternatives selects the destination yielding maximum benefit with a probability equal to one; other destinations having a zero probability of being selected. That is,

$$p_{ij'} = \begin{cases} 1 \text{ if } U_{ij'} > U_{ij} \text{ (for all } j \in N, j \neq j') \\ 0 \text{ otherwise} \end{cases} \tag{4.4}$$

where N represents the set of all possible alternatives. Since an individual's exact utility is unknown, the probability of j' being selected is:

$$p_{ij'} = p[U_{ij'} > U_{ij} \text{ (for all } j \in N, j \neq j')] . \tag{4.5}$$

McFadden (1974a; 1974b) has shown that substituting (4.3) into (4.5), and recognising that $\mu_{ij}$ is a random term drawn from a continuous distribution, yields the traditional general choice model:

$$p_{ij'} = \int_{x=-\infty}^{+\infty} g(\mu_{ij'} = x) \prod_{\substack{j \\ j \neq j'}}^{n} \int_{y=-\infty}^{V_{ij'}-V_{ij}+x} g(\mu_{ij} = y) .dy.dx \tag{4.6}$$

where g( ) represents a probability density function. Since the distribution of the $\mu_{ij}$s is unknown, some continuous distribution has to be assumed. If it is assumed that $g(\mu_{ij})$ is distributed according to a Type I Extreme Value distribution (Fisher and Tippett, 1928), the multinomial logit model arises,

$$p_{ij'} = \exp(V_{ij'})/\Sigma_j \exp(V_{ij}) \tag{4.7}$$

which clearly forms the basis for both the production-constrained interaction model and the attraction-constrained model. For instance, if $V_{ij}$ is defined as,

$$V_{ij} = \sum_h \alpha_h \ln x_{ijh} \qquad (4.8)$$

where x represents a variable describing the attractiveness of a destination, the resulting model form is equivalent to that in equation (2.21). Use of this formulation is reinforced by its intuitively appealing representation of the behavioural response of individuals to changing levels of measurable utility. From equation (4.7),

$$\partial p_{ij}/\partial V_{ij'} = p_{ij'}(1 - p_{ij'}) \qquad (4.9)$$

which yields the logistic curve described in Figure 4.2. The relationship is such that the probability of an individual selecting alternative j' is low when $V_{ij'}$ is low and remains low until $V_{ij'}$ approaches max ($V_{ij}$ j∈ n, j≠j') when it increases rapidly. As $V_{ij'}$ continues to increase beyond the maximum of all other $V_{ij}$ values, $p_{ij'}$ continues to increase but at a decreasing rate and slowly approaches a value of unity. Because of its popularity, the application of the logit model to the analysis of spatial choice is now examined further and, in particular, some problems with its application that stem from the behavioural foundations of the model are discussed.

## 4.7  Problems with the Application of the Logit Formulation to Spatial Choice

At least three potential problems can be identified with the use of the multinomial logit framework to model discrete choice and, in particular, spatial choice. One of these problems pertains to the behavioural assumptions embedded in the derivation of the model, and two pertain to the model's structure, (although model structure is clearly related to the derivation of the model). The potential behavioural problem concerns the assumption that individuals employ an optimising information-processing strategy to arrive at a spatial choice. The two structural problems concern the Independence from Irrelevant Alternatives (IIA) property, and the regularity problem. Each of these problems is now discussed in more detail in the context of spatial choice.

Figure 4.2  **Relationship between $p_{ij}$' and $V_{ij}$' in the Logit Model**

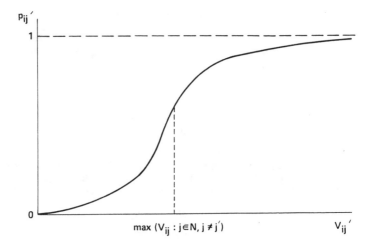

## 4.7.1   The Optimising Information-Processing Assumption

The derivation of the logit framework depends on the assumption that an individual evaluates all possible alternatives which, in a spatial choice context where the number of alternatives is usually large, seems highly implausible. Several authors have recognised that individuals have a limited capacity for processing information (inter alia Simon, 1969; Lindsay and Norman, 1972; Newell and Simon, 1972; Norman and Bobrow, 1975) and that we need some simplifying procedure to make sense of what is observed. Consider, for example, the number of alternatives an individual faces making a residential choice in a large urban area, or a migrant faces in selecting a destination. In both cases it would normally be impossible to evaluate each alternative in accordance with the optimising strategy. Instead, it is likely that an elimination procedure takes place so that only a subset of the original alternatives is evaluated; all others being discarded according to some general criterion or criteria. This type of information-processing efficiency has long been recognised in psychology (Restle, 1961; Sheridan et al., 1975) and in brand choice (Alexis et al., 1968; Lussier and Olshavsky, 1974; Lehtinen, 1974; Rao and Sabavala, 1981; Moore et al., 1985).

It has been suggested in the brand choice literature that a hierarchial choice process operates when the number of alternatives is as low as six (Bettman, 1979, p.221). In most spatial choice situations the number of alternatives far exceeds this threshold which suggests that the predominant information-processing strategy is likely to be a hierarchial one. Consider, for example, the residential choice problem with an individual selecting a house in a city with two hundred suitable vacancies. Clearly, it is unlikely that every vacancy can be evaluated: the individual is thus likely to select a cluster of residences in one part of the city in which to concentrate his/her search. Empirical evidence for such a process has recently been provided by Huff (1986).

The rationale for evaluating clusters of alternatives is summarised in Figure 4.3 which describes the relationship between the number of evaluations necessary to reach a decision and the number of alternatives. When the number of alternatives is small, less than $x_1$, individuals are able to apply an optimising information-processing strategy to reach a decision. As the number of alternatives increases to $x_1$, however, the limit of an individual's information-processing ability is reached and beyond $x_1$ some modification to the processing strategy must be initiated such as dividing the set of alternatives into two clusters and examining only the alternatives in one of these clusters. In a residential choice context, for example, an individual may concentrate his/her search in the western half of the city. As the number of alternatives in the system continues to increase, half of this increase can be expected to occur in the selected cluster so that the number of evaluations necessary to reach a decision rises more slowly between $x_1$ and $x_2$. When the limit of information-processing is again reached, however, at $x_2$ a further simplifying strategy needs to be initiated. In the residential choice example, this may mean dividing each of the two clusters so that only alternatives in one quadrant of the city are evaluated. The process continues until there are too many clusters to evaluate and a third tier in the choice process needs to be developed.

It would thus appear that in spatial choice situations, where the number of alternatives is generally large, individuals are more likely to utilise a hierarchial information-processing strategy than an optimising strategy and the use of the logit formulation is severely limited.

## 4.7.2   The IIA Property in Spatial Choice

The basic implication of the IIA property inherent in the logit formulation can be stated thus: the ratio of the probabilities of an individual selecting two alternatives is unaffected by the addition of a third alternative. While this may be reasonably representative of certain aspatial choice situations, it is very unlikely to occur in spatial choice because of the fixed locations of spatial alternatives. To see this, consider again the system in Figure 4.1 and assume that the alternatives are supermarkets. Suppose that the ratio of the probability that individual i selects supermarket 4 to supermarket 6 is three to one in reality (that is, a consumer is three times as likely to select store 4 than to select store 6). If a new supermarket is built immediately adjacent to store 4, then because of differential competition effects, it would seem logical to assume that the new store would reduce the probability of selecting store 4 by a greater amount than that of selecting store 6. Such a change cannot be modelled with the logit framework because of the IIA property.

### Figure 4.3  A Demonstration of Hierarchical Information Processing

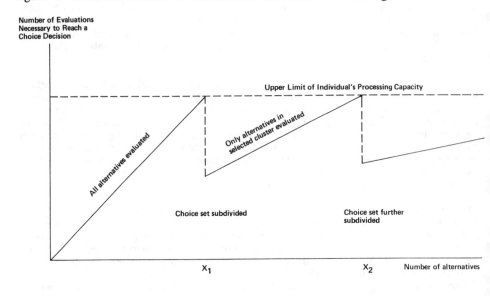

Number of Evaluations
Necessary to Reach a
Choice Decision

Upper Limit of Individual's Processing Capacity

Only alternatives in selected cluster evaluated

All alternatives evaluated

Choice set subdivided

Choice set further subdivided

$x_1$          $x_2$     Number of alternatives

### 4.7.3  The Regularity Property in Spatial Choice

Another property of the logit model that is particularly undesirable in spatial choice is what Huber et al. (1982) term regularity. That is, in the logit formulation it is impossible to increase the probability of selecting an existing alternative by the addition of an alternative to the choice set [the numerator of equation (4.7) remains constant while the denominator increases]. In a spatial choice context such a situation may well occur in reality: an example being the addition to a shopping mall of a retail outlet that increases the custom at existing stores within the mall.

### 4.8  The Nested Logit Approach To Spatial Choice

Given that some or all of the above problems may exist in the use of a logit model to analyze and predict spatial choice, it is hardly surprising that efforts have been made to develop alternative frameworks. One of these, the nested logit model, can be derived from the tenets of the random utility framework described above by assuming that individuals process information hierarchically and that the form of the hierarchy is known to the modeller. Technically, the error terms in equation (4.6) are assumed to follow a standard Generalised Extreme Value distribution (McFadden, 1978; 1980; Williams and Ortuzar,

1982) which necessitates the a priori identification of a set of clusters of alternatives. Individuals are assumed to evaluate each cluster prior to selecting one particular cluster and only within the selected cluster are individual alternatives evaluated. The model resulting from these assumptions can be written in terms of clusters of alternatives each denoted by s in a series of equations:

The probability that an individual i will select a particular spatial cluster s' is,

$$p_{is'} = \frac{\exp(V_{is'})[\sum_{j \in s'} \exp(V_{ij})]^\sigma}{\sum_s \exp(V_{is})[\sum_{j \in s} \exp(V_{ij})]^\sigma} \tag{4.10}$$

the probability that an individual i will select a particular alternative j' within the chosen cluster s' is,

$$p_{ij' \in s'} = \exp(V_{ij' \in s'}) / \sum_{j \in s'} \exp(V_{ij}) \text{ for } j' \in s' \tag{4.11}$$

and the probability of selecting j' from the set of all alternatives is then,

$$p_{ij'} = p_{is'} \cdot p_{ij' \in s'} \tag{4.12}$$

The expression $\sum_{j \in s} \exp(V_{ij})$ is known as an inclusive value and it describes the attractiveness of a cluster resulting from the individual alternatives within that cluster. Its formulation is justified by Williams (1977). The parameter $\sigma$ reflects the extent to which individuals process information hierarchically. If $\sigma = 1$, individuals do not make spatial choices hierarchically but instead evaluate all alternatives prior to making a choice. Under such circumstances, the nested logit and logit formulations are equivalent. If $0 \leq \sigma < 1$, the choice process is to some extent hierarchical. Values of $\sigma$ outside the range 0-1 have no rational interpretation under the tenets of random utility maximisation.

The nested logit model would thus appear to offer a solution to some of the problems with the logit model outlined above and its use is recommended wherever choice is likely to result from a hierarchical information-processing strategy and when the cluster of alternatives perceived by individuals is known to the modeller. In aspatial choice, for example, such clusters are often easily identified and include the choice of regular versus diet brands of soft drink and regular versus decaffeinated brands of coffee. In spatial choice, use of the nested logit would be appropriate in modelling the choice of clothing store where the first choice is that of a shopping mall and the second choice is that of a particular store within the selected mall.

However, use of the nested logit model is likely to be more limited in spatial choice than it is in aspatial choice because frequently the cluster of spatial alternatives perceived by individuals are not discrete but are "fuzzy" (Zadeh, 1965). Consider for example, a consumer's mental map of grocery stores in a large city where there are no obvious clusters of stores. The consumer is likely to have an impression about different shopping neighbourhoods but where the boundaries of such neighbourhoods occur may not be at all obvious to the modeller. The fuzziness of space creates several problems for the application of a nested logit model to spatial choice:

(i) The membership of each cluster of alternatives perceived by an individual cannot be identified precisely by the modeller which creates difficulty in applying the nested logit model. Cluster membership would have to be 'guessed' by the modeller in order to calibrate equations (4.10) and (4.11).

(ii) Individuals in different locations are likely to have different mental maps of the spatial distribution of opportunities and of the way in which alternatives are arranged in clusters. This implies that the modeller would have to guess not just at one set of clusters but at a series of clusters for individuals in different locations.

(iii) Even if the modeller were able to guess accurately at the cluster of alternatives envisaged by different individuals, the nested logit formulation still assumes that the discriminating variable between the clusters is discrete, as in the case of diet versus regular soft drinks, for example, whereas in reality the discriminating variable between spatial clusters is continuous. The inappropriate application of a nested logit model to spatial choice could lead to unrealistic situations such as an alternative near the border of one cluster and an alternative near the border of an adjacent cluster not being considered substitutes for one another yet being located very close together. Similarly, alternatives within the same cluster would, unrealistically in spatial choice, be treated as equal substitutes for one another but yet would be located at varying distances apart.

Because of the above criticisms, it is clear that the nested logit model is only applicable to choice situations where the composition of choice clusters perceived by individuals is known to the modeller. While there are some types of spatial choice in which such information is available, in most instances, spatial clusters are not readily identifiable and the application of the nested logit would involve a great deal of subjectivity and probable error.

## 4.9   The Competing Destinations Approach to Spatial Choice

While the logit and nested logit approaches to modelling discrete choice have, for the most part, been developed and tested in aspatial choice situations, the competing destinations model has been derived from purely spatial considerations (Fotheringham, 1983a; 1986). It thus provides a way of overcoming some of the problems outlined above with the logit and nested logit models that arise from the transference of essentially aspatial theory to the spatial realm.

Three assumptions are made in the derivation of the competing destinations model. One is that there is a limit to an individual's ability to process large amounts of information and therefore spatial choice is likely to result from a hierarchical information-processing strategy whereby a cluster of alternatives is first selected. The second is that due to the continuous nature of space, the composition of spatial clusters perceived by individuals is often unknown to the modeller and so a probability of cluster membership has to be attached to each alternative. The third is that the error terms in equation (4.6) are distributed according to a Type I Extreme Value distribution.

Formally, consider that an individual evaluates only a subset M of the set N of all alternatives. Denote the probability of an alternative $j$ being in the set M selected by individual i as $p_i$ ($j \in M$). Then, in equation (4.5) of the random utility maximisation framework, each alternative's "score" should be weighted by the probability of that alternative being in the restricted choice set in the following way:

$$p_{ij'} = p[U_{ij'} > U_{ij} + \ln p_i (j \in M) \text{ (for all } j \in N, j=j')].p(j' \in M) \tag{4.13}$$

which results in the more general spatial choice model:

$$p_{ij'} = p_i(j' \in M) \int_{x=-\infty}^{+\infty} g(\mu_{ij'}=x) \prod_{\substack{j \\ j \neq j'}}^{n} \int_{y=-\infty}^{V_{ij'}-V_{ij}+x-\ln p_i(j \in M)} g(\mu_{ij}=y).dy.dx \tag{4.14}$$

(Fotheringham, 1988b). Under the assumption that the $\mu_{ij}$s are independently and identically distributed with a Type I Extreme Value distribution, the resulting choice model has the form:

$$p_{ij'} = \frac{\exp(V_{ij'}).p_i(j'\in m)}{\sum_j \exp(V_{ij}).p_i(j\in m)}$$

(4.15)

or, equivalently,

$$p_{ij'} = \frac{\exp(V_{ij'}).l_i(j'\in m)}{\sum_j \exp(V_{ij}).l_i(j\in m)}$$

(4.16)

where $l_i$ ($j \in M$) is the likelihood that an individual i perceives alternative j to be in M.

The choice model described in equation (4.15) is a logit formulation where each alternative's utility function is weighted by the probability of that alternative being evaluated by an individual. It is known in the spatial choice literature as a competing destinations model (Fotheringham, 1983a; 1986).

Both the logit and nested logit models can be obtained as special cases of equation (4.15). If an individual evaluates all possible alternatives, $p_i$ ($j \in M$) = 1 for all $j \in N$, and equation (4.15) is equivalent to the logit model; if an individual processes information hierarchically and choice set membership is known so that $p_i$ ($j \in M$) = 1 for all $j \in M$ and $p_i$ ($j \in M$) = O for all $j \notin M$, equation (4.15) is equivalent to the nested logit model.

By incorporating a weight on each observable utility, the structure of the competing destinations model is intrinsically different from that of the logit model. This is evidenced by the fact that the former no longer contains the undesirable IIA property. The ratio of the probabilities of selecting two alternatives, 1 and 2, from the competing destinations model is

$$p_{i1}/p_{i2} = \frac{\exp(V_{i1}).l_i(1\in m)}{\exp(V_{i2}).l_i(2\in m)}$$

(4.17)

which is no longer necessarily constant under the addition of a new alternative having differential effects on the values of $l_i$ ($1 \in M$) and $l_i(2 \in M)$. Define $L_{12}$ as the ratio of $l_i(1 \in M)$ to $l_i(2 \in M)$ and $L_{12*}$ as the value of $L_{12}$ after the addition of a new alternative. Then, as Batsell and Polking (1985) note, three situations are possible: (i) $L_{12*} > L_{12}$ indicating that the new alternative increases the relative attractiveness of 1 over 2; (ii) $L_{12*} = L_{12}$ indicating that the addition of the new alternative has no effect on the relative attractiveness of the existing alternatives (the logit assumption); and (iii) $L_{12*} < L_{12}$ indicating that the new alternative increases the relative attractiveness of 2 over 1.

The other undesirable property of the logit model, regularity, is also no longer present in the competing destinations model. It is possible to replicate an increase in the probability of selecting an existing alternative due to the addition of a new alternative wherever the increase in $l_i(j \in M)$ due to the addition of the new alternative is greater than the increase in the denominator of equation (4.16).

Standard maximum likelihood or least squares techniques can be employed to calibrate the competing destinations model, the only added complexity being the necessity to define $l_i(j \in M)$, the likelihood that an alternative is in the restricted choice set. Two closely related approaches to measuring $l_i(j \in M)$ presently exist. One approach considers that the likelihood of a particular alternative being in the restricted choice set is a function of the dissimilarity of that alternative to all others (Batsell, 1981; Meyer and Eagle, 1982; Borgers and Timmermans, 1987). The rationale for this approach is that the degree to which an alternative possesses distinctive properties affects its chances of being included in M. Whether it is affected positively or negatively is an empirical question.

The other approach to measuring $l_i(j \in M)$ is more applicable to spatial choice and recognises that the more proximal are alternatives in geographic space, the more likely they are to be substitutes for one another. Hence, the location of an alternative with respect to all other alternatives affects its chances of being included in the restricted choice set of an individual. Fotheringham (1983) suggests measuring $l_i(j \in M)$ in this way by using a sum of weighted distances from one alternative to all others where the weight is the size of each alternative. That is,

$$l_i(j' \in M) = \left( (1/n-1) \sum_{j \neq j'} w_j/d_{j'j} \right)^\theta \qquad (4.18)$$

where $w_j$ represents the weight of alternative j and $d_{j'j}$ represents the distance between the two alternatives j' and j. This is a traditional measure of relative location within disciplines concerned with the role of space in human activities and is sometimes referred to as potential accessibility (Hansen, 1959). Large values of the variable indicate alternatives that are in close proximity to other alternatives; low values indicate alternatives that are spatially isolated.

Using this definition of $l_i(j \in M)$ and defining c as the measure of centrality, the competing destinations model can be written as;

$$p_{ij} = \frac{\exp(V_{ij}).c_{j'}^{\theta}}{\sum_j \exp(V_{ij}).c_j^{\theta}} \qquad (4.19)$$

whence it is clear that if $\theta = 0$, equation (4.19) is equivalent to the logit formulation indicating that information is not being processed hierarchically and that all alternatives are evaluated simultaneously. Thus, the degree to which $\theta$ diverges from zero, which can be examined by a t-test or a relative-likelihood statistic, indicates the degree to which information is processed hierarchically.

The sign of $\theta$ can be positive or negative depending on the empirical context in which the choice model is calibrated. It depends on the intrinsic attractiveness of large clusters of alternatives in a manner described more fully elsewhere (Fotheringham, 1986). If the attraction of a cluster increases exponentially as the number of alternatives in it increases, $\theta$ will be positive reflecting some sort of agglomeration and relationship whereby the closer is j' to other alternatives, the more likely it is to be selected, ceteris paribus. Conversely, if the attraction of a cluster increases logarithmically with its size, $\theta$ will be negative reflecting some sort of competition relationship whereby alternatives in close proximity to others are less likely to be selected than peripheral ones, ceteris paribus. The sign of $\theta$ is a matter for further empirical research and may well vary in different choice situations. An example of the estimation of $\theta$ in a migration context is provided in Chapter 5.

## 4.10 The Misspecification of Spatial Interaction Models: Implications for Predicting Flows

It has been postulated that hierarchical choice is an efficient way (and sometimes the only way) of processing large amounts of information. Individuals making residential choices or migrants making destination choices are unlikely to be able to evaluate all the possible alternatives available to them; the decision process must proceed so that a manageable number of evaluations are made. The obvious way to process a large number of alternatives efficiently is to exclude a large number of them based on some attribute and evaluate only the remaining alternatives. In a spatial choice context we assume this attribute is related to location so that many destinations are never evaluated, not because of their individual characteristics, but because of their general location. It is assumed that individuals have positive or negative attitudes towards regions which allows them to concentrate their search over a limited number of locations. Whether individuals identify one particular region to which they would like to move or they identify one or more regions to which they would not like to move, they are reducing the number of evaluations that are necessary to reach a decision on a particular destination and are, hence, making their spatial choice hierarchically.

It is useful to consider errors that can arise in modelling spatial choices that result from a hierarchical process with models, such as the singly constrained interaction model, that are based on the assumption of an optimising information-processing strategy. For this purpose, assume, for the sake of exposition, that the observable utility an individual would receive from selecting a particular spatial alternative is constant across alternatives. That is,

$$V_{ij} = V_i \text{ for all } j \tag{4.20}$$

Let $n_s$ represent the number of outlets in cluster s and $V_{is}$ be the perceived attractiveness of cluster s. Consider the two information-processing strategies, optimising and hierarchical, described above.

If individuals evaluate all alternatives without regard to their spatial clustering, the attractiveness of a spatial cluster of alternatives is merely the sum of the attractions of the individual alternatives within that cluster. That is,

$$V_{is} = \sum_{j\in s} V_{ij} \tag{4.21}$$

which on substituting equation (4.20) implies that

$$V_{is} = n_s V_i \tag{4.22}$$

Consider now what happens to the perceived attractiveness of the cluster as $n_s$ increases. The rate of increase in this attractiveness is,

$$\partial V_{is}/\partial n_s = V_i \tag{4.23}$$

which is a constant and is represented by the slope a in Figure 4.4. That is, regardless of the original size of a cluster, the addition of a new alternative to the cluster will increase its perceived attractiveness by $V_i$, the perceived attractiveness of the new alternative. In many circumstances of spatial choice it would seem unreasonable to expect such a relationship to occur in reality. For example, the addition of one store to a cluster of 100 stores probably does not add as much to the perceived attractiveness of the cluster as does the addition of the same store to a cluster of 25. Alternatively, a cluster of 25 stores may be perceived as being more than 25 times as attractive as an individual store. In either case, such relationships cannot be modelled with the logit formulation where the relationship between perceived attractiveness and size is described by slope a.

Suppose, however, that individuals first evaluate clusters of alternatives prior to evaluating the alternatives within a chosen cluster. That is, assume individuals process spatial information hierarchically. Then, the perceived attractiveness of a cluster can be described by,

$$V_{is} = \alpha f(\textstyle\sum_{j\in s} V_{ij}) \tag{4.24}$$

where f( ) is some functional form to be determined theoretically or empirically and $\alpha$ is a parameter relating $V_{is}$ to $V_{ij}$. In what follows it is necessary to select a specific functional form and, for convenience, assume that the perceived attractiveness of a cluster and the sum of its constituents are related through a power function. That is,

$$V_{is} = (\textstyle\sum_{j\in s} V_{ij})^{\alpha} \tag{4.25}$$

Clearly equation (4.25) is a more general statement of the relationship between these two variables than is equation (4.21) in which $\alpha$ is assumed to equal one. Consider now what happens to $V_{is}$ as the size of the cluster increases. Noting equation (4.20), equation (4.25) can be rewritten as,

$$V_{is} = n_s^{\alpha} V_i^{\alpha} \tag{4.26}$$

so that,

$$\partial V_{is}/\partial n_s = \alpha n_s^{\alpha-1} V_i^{\alpha} \tag{4.27}$$

which depends critically on $\alpha$ in the following manner:

$$\text{as } n_s \rightarrow 0, \ \partial V_{is}/\partial n_s \rightarrow \begin{cases} \infty & \text{if } 0 < \alpha < 1 \\ V_i & \text{if } \alpha = 1 \\ 0 & \text{if } \alpha > 1 \end{cases}$$

$$\text{as } n_s \rightarrow \infty, \ \partial V_{is}/\partial n_s \rightarrow \begin{cases} 0 & \text{if } 0 < \alpha < 1 \\ V_i & \text{if } \alpha = 1 \\ \infty & \text{if } \alpha > 1 \end{cases} \tag{4.28}$$

When $0 < \alpha < 1$, these relationships produce slope c in Figure 4.4. The rate at which the perceived attractiveness of a cluster continues to increase as alternatives are added, decreases so that some sort of spatial discounting takes place.

Figure 4.4  **Relationships between the Perceived Attractiveness of a Cluster and the Size of the Cluster**

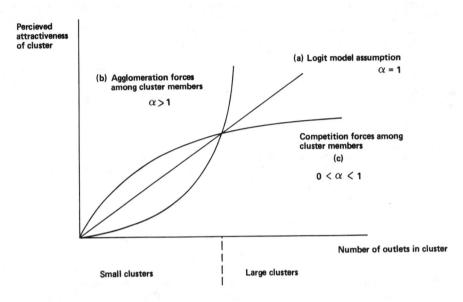

In such a situation there are competitive forces among alternatives since the addition of a new alternative to the cluster will decrease the probability of existing alternatives in the cluster being selected by the individual.

When $\alpha > 1$, slope b in Figure 4.4 results. Here the rate at which the perceived attractiveness of a cluster increases at an increasing rate as alternatives are added to the cluster. In such a situation, agglomeration forces exist among alternatives since the addition of a new alternative to the cluster increases the probability of an individual selecting one of the existing outlets.

When $\alpha = 1$, the relationship between perceived attractiveness and cluster size is that assumed in the logit model. The restrictive nature of such an assumption is clearly seen in Figure 4.4 where the relationship between perceived attractiveness and cluster size represented by the value of $\alpha = 1$ is but one of an infinite number of relationships that could occur. It thus seems highly unrealistic in most spatial choice situations to employ the logit model which is based upon this assumption.

To see the errors that would arise when the logit model is applied to situations where individuals evaluate alternatives hierarchically and when $\alpha$ is not equal to 1, compare the slopes in Figure 4.4. When $\alpha > 1$ and agglomeration forces exist, the logit model would overpredict the probability of choosing alternatives in small clusters and underpredict the probability of choosing alternatives in large clusters.

When $0 < \alpha < 1$ and competition forces exist, the logit model would underpredict the probability of choosing alternatives in small clusters and overpredict the probability of choosing alternatives in large clusters.

## 4.11   The Misspecification of Spatial Interaction Models: Implications for Parameter Estimates

To demonstrate the potential misspecification bias in parameter estimates resulting from the calibration of an interaction model which is based on the logit formulation and assumes a non-hierarchical information-processing strategy on the part of individuals, consider the following general form of the production-constrained model:

$$T_{ij} = f(\mathbf{W}_j, d_{ij}) \tag{4.29}$$

where $\mathbf{W}_j$ represents a vector of destination **site** attributes and $d_{ij}$ represents distance. The equivalent competing destinations model is,

$$T_{ij} = f(\mathbf{W}_j, d_{ij}, c_j) \tag{4.30}$$

where $c_j$ represents a destination **situation** variable defined as the centrality of a destination measured with respect to all competing destinations. Both of these models can be made linear by the transformation described in Section 3.2.2 and can then be written as:

$$T_{ij}^* = k_i^* + \sum_h \alpha_h w_{jh}^* + \beta d_{ij}^* + \in_{ij}^* \tag{4.31}$$

and

$$T_{ij}^* = k_i^* + \sum_h \alpha_h w_{jh}^* + \beta d_{ij}^* + \theta c_j^* + \mu_{ij}^* \tag{4.32}$$

respectively. The notation "*" represents the necessary linear transformation which in the case of a production-constrained model is the logarithm of the variable divided by its geometric mean as shown in Section 3.2.2. The terms $\in_{ij}$ and $\mu_{ij}$ represent random errors. Note that in equations (4.31) and (4.32), and in the subsequent discussion, it is assumed that $\beta$ is negative.

Consider, for ease of exposition, that flows are only a function of one site variable so that the models can be written in their simplest form as:

$$T_{ij}^* = k_i^* + \alpha w_j^* + \beta d_{ij}^* + \in_{ij}^* \tag{4.33}$$

and

$$T_{ij}^* = k_i^* + \alpha w_j^* + \beta d_{ij}^* + \theta c_j^* + \mu_{ij}^* \tag{4.34}$$

respectively. If equation (4.34), the competing destinations formulation, represents the "true" model, then the error term associated with the production-constrained model in equation (4.33) is

$$\in_{ij}^* = \theta c_j^* + \mu_{ij}^* \tag{4.35}$$

Whence, it is well-known (inter alia, Hanushek and Jackson, 1977 pp 79-86) that the expected estimates of the parameters $\alpha$ and $\beta$ in the production-constrained model are

$$E(\alpha') = \alpha + \theta \left[ \frac{r(c^*,w^*) - r(d^*,w^*).r(c^*,d^*)}{1 - r^2(d^*,w^*)} \cdot \frac{s(c^*)}{s(w^*)} \right] \tag{4.36}$$

and

$$E(\beta') = \beta + \theta \left[ \frac{r(c^*,d^*) - r(d^*,w^*).r(c^*,w^*)}{1 - r^2(d^*,w^*)} \cdot \frac{s(c^*)}{s(d^*)} \right]$$  (4.37)

where $r(x,y)$ represents the correlation coefficient between variables $x$ and $y$, $r^2(x,y)$ represents the square of this coefficient, and $s(x)$ represents the standard deviation of variable $x$. Consequently, if equation (4.30) does represent a "true" model of interaction, the right-hand expressions in equations (4.36) and (4.37) represent the misspecification bias present in the estimates of $\alpha$ and $\beta$, respectively, from a traditional production-constrained model. Fotheringham (1984) has shown that these biases can be interpreted in terms of spatial structure. Consider the bias in the distance-decay parameter estimated separately for each origin in a system. From equation (4.37), this will be

$$E(\beta_i') = \beta + \theta_i \left[ \frac{r_i(c^*,d^*) - r_i(d^*,w^*).r_i(c^*,w^*)}{1 - r_i^2(d^*,w^*)} \cdot \frac{s_i(c^*)}{s_i(d^*)} \right]$$  (4.38)

A priori, it can be assumed that the linear relationship between $w^*$ and $d^*$ is zero so that equation (4.38) reduces to:

$$E(\beta_i') = \beta + \theta_i[r_i(c^*,d^*).s_i(c^*)/s_i(d^*)]$$  (4.39)

which is equivalent to,

$$E(\beta_i') = \beta + \theta_i\delta_i'$$  (4.40)

where $\delta_i'$ represents the estimated slope parameter obtained in regressing $c^*$ on $d^*$.

Hence, the bias in $\beta_i'$ obtained from an interaction model without a competing destinations variable is given by the term $\theta_i\delta_i'$ which can be interpreted as the indirect elasticity of interaction with respect to distance: $\theta_i$ is the elasticity of interaction with respect to destination centrality; $\delta_i'$ is the estimated elasticity of destination centrality with respect to distance. Clearly, if $\theta_i = 0$ and/or $\delta_i' = 0$, there is no bias in the estimated parameter. The former situation arises whenever there is no linear relationship between the transformed interaction and centrality variables; the latter arises whenever there is no linear relationship between the transformed functions of centrality and distance.

The direction of the bias in $\beta_i'$, which is equivalent to the sign of the expression $\theta_i\delta_i'$, depends on the location of i as described in Figure 4.5. The sign of $\theta_i$ depends on the prevalence of either economic forces ($\theta_i > 0$) or psychological forces ($\theta_i < 0$). In the former instance, individuals are motivated to select destinations in close proximity to other destinations in order to minimise expected travel beyond the first destination, a behaviour typical in shopping patterns, for example. In the latter instance, individuals are less likely to select destinations in close proximity to other destinations, ceteris paribus, because they underestimate the number of opportunities in large clusters of alternatives. Fotheringham (1988c) provides more detail on these two contrasting determinants of spatial choice and an empirical example of the latter is provided in the subsequent chapter on migration.

Figure 4.5  **An Example of Parameter Bias in an Interaction Model not Containing a Destination Competition Term**

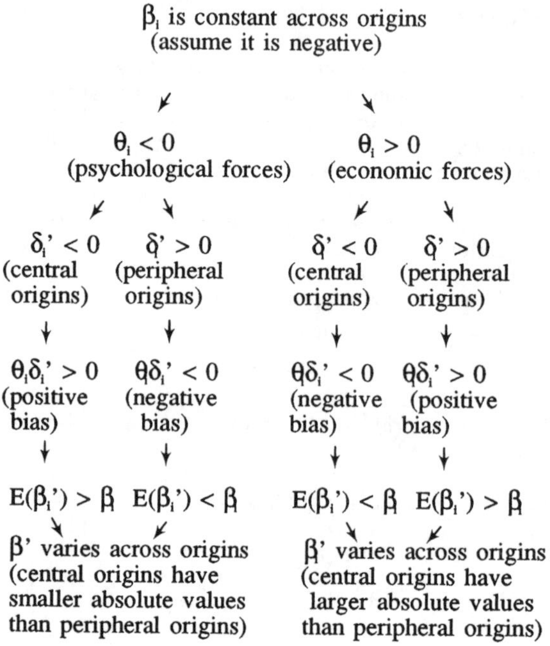

$\beta_i$ is constant across origins
(assume it is negative)

$\theta_i < 0$
(psychological forces)

$\theta_i > 0$
(economic forces)

$\delta_i' < 0$
(central origins)

$\delta_i' > 0$
(peripheral origins)

$\delta_i' < 0$
(central origins)

$\delta_i' > 0$
(peripheral origins)

$\theta_i \delta_i' > 0$
(positive bias)

$\theta_i \delta_i' < 0$
(negative bias)

$\theta_i \delta_i' < 0$
(negative bias)

$\theta_i \delta_i' > 0$
(positive bias)

$E(\beta_i') > \beta_i$    $E(\beta_i') < \beta_i$

$E(\beta_i') < \beta_i$    $E(\beta_i') > \beta_i$

$\beta'$ varies across origins
(central origins have
smaller absolute values
than peripheral origins)

$\beta'$ varies across origins
(central origins have
larger absolute values
than peripheral origins)

Although the sign of $\theta_i$ is a matter for empirical research, it can be assumed that for a particular type of interaction, $\theta_i$ will have the same sign for all origins. The sign of $\delta_i'$, however, will not be constant across all origins. The parameter reflects the relationship between destination centrality and distance from the origin. For a centrally located origin this relationship will be negative; as distance from such an origin increases, centrality will decrease. For a peripheral origin, on the other hand, the relationship will be positive; as distance from such an origin increases, centrality will increase. Hence, the bias in $\beta_i'$ will have a different sign for central and peripheral origins. For example, if $\theta_i$ is negative, $\theta_i \delta_i'$ will be positive for central origins and negative for peripheral origins. The resulting estimated origin-specific distance-decay parameters will then exhibit a marked spatial pattern in which the values for central origins will appear smaller in absolute terms than those for peripheral origins (it is assumed that $\beta_i$ is negative for all origins). As Fotheringham (1981) demonstrates, this type of pattern is often reported in studies where origin-specific interaction models have been calibrated. This pattern, however, appears likely to be a product of model misspecification rather than of some fundamental spatial difference in the behaviour of individuals. More detail on the misspecification bias in parameter estimates obtained in the calibration of interaction models can be found in Fotheringham (1984).

# PART 2

APPLICATIONS

# CHAPTER 5

---

## APPLICATIONS TO MIGRATION ANALYSIS

### 5.1 Introduction

The term 'migration' has been applied to the movements of individuals over space at many different scales, from the international to the intraurban. However, the vast majority of studies in which interaction models have been applied are at the mesoscale of interregional or interurban movements and we restrict our discussion to this level.

The major purpose of this chapter is thus to demonstrate the use of spatial interaction modelling in the analysis of interregional or interurban migration tables. We describe in particular the application of an extended production-constrained model to identify the determinants of the spatial choices of migrants within the United Kingdom. We also use this example, however, to demonstrate some of the points made in made in Chapter 4 regarding the way individuals process spatial information in order to choose a destination. Through the calibration of a competing destinations model we are able to report on the extent of hierarchical information processing and on spatial variations in such processing. These results are reinforced by a similar application to Dutch migration tables. We also discuss two different approaches to obtaining information on migration flows from spatial interaction modelling. The first is the estimation of migrants' psychological distances by Plane (1984) and the subsequent distortion of space produced by such distances; the second is Tobler's (1979) estimation of place attraction through knowledge of the marginal totals in a migration matrix. Both of these studies make use of US migration tables; Plane's study analyses interurban migration while Tobler's analyses interstate migration. However, we first comment on why a large proportion of the spatial interaction modelling literature has been concerned with the application of models to migration and then give a brief sampler of such applications.

### 5.2 The Importance of Understanding Migrants' Destination Choices

Being able to understand, and thereby possibly forecast or affect, the outcomes of a large number of decision processes regarding relocation within a country is an important skill. Apart from satisfying a basic curiosity regarding why and where people move, migration modelling has a long history across many disciplines for several reasons:

(1) In most countries, the internal relocation of individuals is the prime determinant of regional population growth rates. Given the recent decline in the natural growth rate of population, particularly in developed countries, interregional migration plays an ever-increasing role in producing variations in regional population totals. There is usually substantial government interest in such population redistribution because, while a symmetrical redistribution tends to be beneficial as an important mechanism for promoting integration, long-term, asymmetrical migration patterns can result in large population imbalances. In extreme cases, this can lead to political instability and certain regions being overpopulated and essentially insupportable. The recent rapid growth of many primate cities in third world countries is partly the result of such migration asymmetry and has been examined with the aid of spatial interaction models (Ledent, 1982).

(2) Migration has long been thought of as an equilibrating process through which reductions are brought about in regional variations in the supply and demand for labor (Cebula and Vedder, 1973). Pressure can be released on the local economies of depressed

regions by the outmigration of unemployed workers (Vanderkamp, 1971). However, there are conflicting views on such a process because migration tends to be a selective process and generally migration streams contain higher proportions of those most able to find employment than are found in the origins of such streams (Brown and Burrows, 1977). In the long run, this leads to depressed regions losing their most productive workers and leaves them with ever lower participation rates, productivity rates and potential for growth. For these and other reasons, most developed countries have enacted a variety of policies aimed at reducing regional disparities and stemming asymmetrical interregional migration.

(3) In much the same way as described above, migration is strongly linked to regional economic growth and decline (Muth, 1971). In particular, net outmigration from regions can have strong negative multiplier effects on the local economy as shown in Figure 5.1. It is not just the direct loss of migrants that causes regional economic decline but it is also the loss of the income spent in the local economy by those migrants which leads to further job losses in the service sector, creating the potential for a further round of income loss and outmigration. This multiplier effect can also work in reverse so that regions of net inmigration receive the benefits of not only those net inmigrants but also of the extra jobs created by their spending of income in the local economy.

(4) Large scale movements of people within a country produce regional variations in the demand for, and the price of, housing. In the United Kingdom, for example, partly because of the continued attraction of the southeast, house prices in that region are well above the national average while those in the more depressed regions in the north are well below average. Similarly, the large increase in house prices in California during the late 70s and early 80s was fuelled by the attractiveness of that state to migrants from the rest of the country.

Figure 5.1 **The Negative Multiplier Effect of Outmigration**

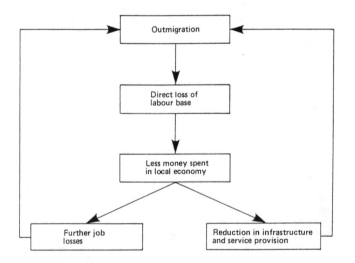

(5) As noted above, large scale movements of population within a country are largely responsible for variations in regional population totals and growth rates. In many countries, such variations are directly related to variations in political power: regions with large populations tend to be dominant in shaping the political views of the country. Nowhere is

this more clearly seen than in the United States where the numbers of representatives elected to Congress from each state are in direct proportion to the state's population. After each census, these numbers are adjusted with some states losing representatives and others gaining (after the 1980 Census, for example, Florida gained five representatives while New York lost three). Similarly, voting in presidential elections and primaries is also by state with the states having the largest populations (California, New York and Texas) generally having a very strong influence in the election of candidates. Haynes and Fotheringham (1984) report an exercise in the use of spatial interaction models to forecast the voting behaviour of a hypothetical state experiencing rapid inmigration from politically diverse sources.

## 5.3  Examples of the Application of Spatial Interaction Models in Migration Analysis

The intention of this and the subsequent 'applications' chapters of the book is to demonstrate the type of analysis in a subject area that can be undertaken with the aid of spatial interaction models. We concentrate on applicability and on results that yield practical and useful information on the system being analysed. In this particular chapter, for example, the questions that are asked are "What are the determinants of migrants' destination choices?" and "To what extent are such choices made in a hierarchical manner?". We therefore intentionally avoid extensive literature reviews which can be found elsewhere in many of the references cited throughout the book. However, it would be unwise to avoid the earlier literature entirely and below we discuss briefly a selection of applications of spatial interaction models to migration. The emphasis is on the utility of the findings and on the type of application commonly found. We discuss two applications in more detail that are interesting in their originality: one is the estimation of psychological distances by Plane (1984); the other is the estimation of destination attractiveness by Tobler (1979).

The investigation of the determinants of migrants' destination selections through the use of spatial interaction models has a very long history. Indeed, the early development of spatial interaction models was prompted by certain regularities in migration streams that came to be recognised by Ravenstein (1876; 1885; 1889) who noted: "the great body of migrants only proceed a short distance" and that in estimating migration flows we "must take into account the number of natives of each country which furnishes the migrants, as also the population of the towns or districts which absorb them" (Ravenstein, 1885 p.12). These observations led to the development of very simply gravity interaction models with the following formula:

$$T_{ij} = P_i \, P_j / d_{ij}^2, \qquad\qquad (5.1)$$

where $P_i$ and $P_j$ represent origin and destination populations, respectively. This formula formed the foundation of all the members of the family of interaction models described in Chapter 2.

Given its early beginnings, the literature on the application of spatial interaction models to uncover the determinants of migration patterns is vast. As Ewing (1976) notes:

> The mathematical modeling and prediction of interregional
> migration has been one of the more extensively researched
> themes in spatial interaction analysis.

Apart from its early start, there are two main reasons for the widespread use of spatial interaction models in migration. One is the importance of migration as discussed above; the other is the availability of migration matrices (see Haynes and Fotheringham, 1984 for a review of data availability). The latter, in particular, has led to many theoretical advances in spatial interaction modelling being tested with migration data.

Applications of spatial interaction models to migration can broadly be divided into three categories along the lines of the division of spatial interaction models developed in Chapter 2. Information on various aspects of the determinants of migration patterns has been obtained from unconstrained, singly constrained and doubly constrained models. Unconstrained models, despite their drawbacks in terms of internal consistency and accuracy identified in Chapters 2 and 3, have been employed to examine the relative roles of origin and destination characteristics in determining migration patterns. Examples of the use of such models can be found in many different spatial systems including the United States (Blanco, 1963; Greenwood and Sweetland, 1972), Sweden (Isbell, 1944), India (Greenwood, 1971), Canada (Laber and Chase, 1971; Lycan, 1969), the United Kingdom (Flowerdew and Salt, 1979), and Egypt (Greenwood, 1969). Similar studies can be found on outmigration from Appalachia (Clark and Ballard, 1980), migration within California (Rogers, 1967) and the movement of Edinburgh apprentices in the 17th and 18th centuries (Lovett, Whyte and Whyte, 1985).

A typical set of results from the application of an unconstrained spatial interaction model is that provided by Kau and Sirmans (1979) and repeated here in Table 5.1. Kau and Sirmans examine the influence of several origin and destination characteristics on migration patterns for US interstate migration in four time periods, 1940, 1950, 1960 and 1970. Their migration model is:

$$\ln T_{ij}^{t} = \alpha_0 + \alpha_1 \ln y_i + \alpha_2 \ln y_j + \alpha_3 \ln a_i + \alpha_4 \ln e_i + \alpha_5 \ln c_i +$$

$$\alpha_6 \ln c_j + \beta_1 \ln d_{ij} + \beta_2 \ln T_{ij}^{t-1} \qquad (5.2)$$

where $T_{ij}^{t}$ represents the migration from i to j in time period t, y represents average income, a represents average age level, e represents average level of education attained, c represents a climate variable measured by average daily temperature, $d_{ij}$ represents distance and $T_{ij}^{t-1}$ represents the migration from i to j in the previous time period.

As can be seen from Table 5.1, migrants appear to be repulsed by low incomes and high temperatures at an origin and attracted by high incomes and low temperatures at a destination. Fewer individuals appear to migrate from areas having high concentrations of elderly and less well educated individuals (which is evidence that migration tends to be selective and practiced more by the younger and well-educated). Migrants appear to be deterred by distance and attracted by destinations which have been selected by migrants in previous time periods. Notice that through time migrants appear to be less constrained by distance. However, before accepting these results, several cautionary comments should be made. First, as discussed in Chapter 2, the unconstrained model is a relatively inaccurate representation of the underlying migration trends and this inaccuracy casts doubt on the validity of the parameter estimates obtained in the calibration of such a model. Secondly, this particular model does not contain a destination competition variable and so its parameter estimates are likely to contain a misspecification bias as discussed in Chapter 4. Thirdly, the use of the 'migrant stock' variable, $T_{ij}^{t-1}$, is highly dubious in that it is likely to lead to model circularity: flows in one period are 'explained' in terms of flows in a previous period while what has produced the flow pattern in the previous time period is not determined. It is obvious that $T_{ij}^{t}$ and $T_{ij}^{t-1}$ are likely to be highly correlated and therefore that $T_{ij}^{t-1}$ is likely to be highly correlated with the other independent variables in the model so producing a further problem in interpreting parameter estimates (see Fotheringham, 1983a). Fourthly, the climatic variable, temperature, would appear to be a surrogate for some other regional influence. It is unlikely that individuals are in fact attracted by lower temperatures. More likely, this variable simply reflects a trend during this period of migrants moving from the south to the north in search of employment; a trend that changed rapidly during the 1970s and 1980s. Finally, Kau and Sirmans excluded from the analysis all flows that were zero; an extremely dubious procedure for handling zero interactions (see Section 3.3.3). We later report a study of UK migration flows in which all of these problems are absent.

The use of unconstrained interaction models has now been largely superseded in migration analysis by the use of singly constrained models. Production-constrained models (Section 2.3.2) are employed to determine the attributes of destinations affecting choice by migrants while attraction-constrained models (Section 2.3.3) are employed to identify the attributes of origins affecting variations in the propensity of outmigration. While relatively few studies of the latter exist (although see Fotheringham and Flowerdew, 1988, for an exception), there are many examples of the former (inter alia, Ewing, 1976; Fotheringham, 1987; 1988c; Ishikawa, 1987; and Opt'Veld, Bijlsma and Starmans, 1984). In subsequent sections we report two studies of this type in the analysis of migration flows in the United Kingdom and in the Netherlands.

Table 5.1 **Kau and Sirmans' Nonrecursive Migration Results**

| Variable | 1970 | Parameter 1960 | Estimates 1950 | 1940 |
|---|---|---|---|---|
| $y_i$ | -1.54 (6.69) | -3.54 (9.33) | -2.78 (8.13) | -2.24 (13.87) |
| $y_j$ | 1.96 (11.64) | 1.63 (12.87) | 0.86 (7.01) | 1.18 (15.38) |
| $a_i$ | -3.09 (8.68) | 0.11 (0.22) | -0.19 (0.32) | 0.52 (1.08) |
| $e_i$ | 7.10 (13.45) | 5.14 (17.92) | 8.11 (24.67) | 10.96 (30.48) |
| $c_i$ | 0.13 (4.39) | 0.13 (4.71) | 0.12 (3.86) | 0.01 (1.89) |
| $c_j$ | -0.39 (15.48) | -0.50 (18.54) | -0.50 (16.61) | -0.51 (17.06) |
| $d_{ij}$ | -0.38 (12.49) | -0.48 (16.18) | -0.49 (14.97) | -0.72 (20.72) |
| $T_{ij}^{t-1}$ | 0.45 (32.74) | 0.43 (34.01) | 0.42 (31.74) | 0.47 (30.61) |

Figures in brackets represent t statistics

Finally, use of the doubly constrained model (Section 2.3.4) has generally been restricted to three purposes: one is to comment on variations in distance-decay parameters (Stillwell, 1978); a second is to construct forecasts of migration patterns under various projections of the marginal totals (Hallefjord and Jornsten, 1985); a third is to provide insight into the migration process via analytical manipulation. (Below we provide two examples of this last use of the doubly constrained model in migration.) Stillwell's study of variations in distance-decay parameters yields some interesting information on the propensity of individuals to migrate over short distances rather than long distances. Calibration of a doubly constrained model with data on 1961-66 migration between counties in England and Wales yielded distance-decay parameters of 1.2822 (power) and 0.0112 (exponential). For comparison, Plane's (1984) study of US interurban migration 1975-76 which yielded comparable parameters of 1.582 and 0.002295, respectively. Two comments are useful here. First, it is clear that as mentioned in Chapter 1, the power function parameter has the attractive feature of being scale-independent whereas the exponential is not. The difference between the exponential parameters suggests the average migration length in the US in approximately five times that in the United Kingdom. Second, the difference between the power function parameters suggests that, after accounting for the difference in scale between the two countries, British migrants are less deterred by distance than are their US counterparts. Stillwell disaggregated migrants by origin and by age and sex. The origin disaggregation yields information on the spatial variation of the distance-decay parameters which range from -0.7924 (Bedfordshire) to 2.8832 (mid-Wales) when a power function was employed. More comment will be made on the spatial variation of such parameters subsequently. Finally, the age/sex disaggregation produced twenty-four groups of individuals for whom the power distance-decay parameter estimates are given in Figure 5.2 which is reproduced from Stillwell (1978). These results suggest that the propensity to migrate over long distances is at a maximum during the ages twenty to forty, presumably as people search for jobs, and is at a minimum in old age (which are findings supported by Rossi, 1955 and Lowry, 1966). In all age groups except the elderly it appears that males have a greater propensity to migrate long distances than do females, again a finding perhaps related to job-searching.

Figure 5.2 **Variations in Distance-Decay Parameters by Age/Sex for British Migrants (after Stillwell, 1978)**

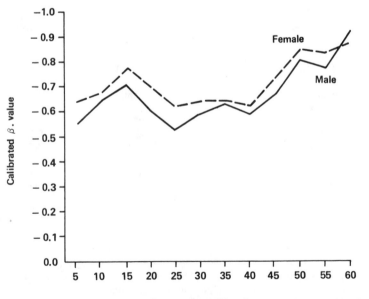

## 5.4  Plane's Derivation of Migrants' Psychological Distances

Consider a doubly constrained interaction model with a power function of distance,

$$T_{ij} = A_i O_i B_j D_j d_{ij}^{-\beta} \epsilon_{ij} \tag{5.3}$$

where $T_{ij}$ is the observed flow and $\epsilon_{ij}$ is an error term representing the difference between $T_{ij}$ and $A_i O_i B_j D_j d_{ij}^{-\beta}$. Plane (1984) suggested turning this equation around and solving for $d_{ij}$ with a known set of $T_{ij}$s. In essence, he assumes that any errors resulting in equation (5.3) are due to an incorrect definition of distance. As discussed in Chapter 2, accurate measurements of the distances that affect an individual's choice of destination are very difficult to obtain since in many cases individuals act on a cognised environment rather than on an objective environment. Plane's idea then was to try to obtain information of the cognitive environment of US migrants. Solving (5.3) for $d_{ij}$ yields

$$\ln d_{ij} = \beta^{-1} \ln A_i + \beta^{-1} \ln B_j + \beta^{-1} \ln(O_i D_j / T_{ij}) + \beta^{-1} \ln \epsilon_{ij} \tag{5.4}$$

which is linear in parameters and can be calibrated by regression. A set of inferred distances can then be obtained by calculating $\exp(\ln d_{ij})$ for each i-j pair. A problem, however, with this strategy is that the balancing factors, $A_i$ and $B_j$, are unknown. Plane solves this problem by noting that the $A_i$s and $B_j$s can be replaced by a series of (m-1) and (n-1) dummy variables, respectively, where m is the number of origins and n is the number of destinations. Hence, the model to be calibrated is:

$$\ln d_{ij} = \alpha_0 + \sum_{k=1}^{m-1} \alpha_k A_k + \sum_{h=1}^{n-1} \tau_h B_h + \sigma \ln(O_i D_j / T_{ij}) + \mu_{ij} \tag{5.5}$$

where,

$$A_k = \begin{cases} 1 \text{ if } k = i \\ 0 \text{ otherwise,} \end{cases} \tag{5.6}$$

$$B_h = \begin{cases} 1 \text{ if } h = j \\ 0 \text{ otherwise,} \end{cases} \tag{5.7}$$

and $\mu_{ij}$ is an error term. The original balancing factors can be recovered from the above equations by,

$$A_i = \exp((\alpha_0 + \alpha_i)/\sigma) \quad (i = 1,...,m-1) \tag{5.8}$$

$$A_m = \exp(\alpha_0/\sigma) \tag{5.9}$$

$$B_j = \exp(\tau_j/\sigma) \quad (j = 1,...,n-1) \tag{5.10}$$

$$B_n = 1 \tag{5.11}$$

and the distance-decay parameter is simply $\sigma^{-1}$.

Plane provides examples of estimating $\ln d_{ij}$ from two sets (1975-76 and 1978-79) of interstate migration flows in the US. The flows were restricted to those between the 48 contiguous states. The two estimated distance-decay parameters were 1.582 and 1.623

respectively. Maps of "perceived space" were then drawn for various origins and destinations. Only sets of distances focussed on one origin or one destination can be displayed graphically because there is no guarantee that perceived distances have metric properties. States having the most "warped" psychological surfaces were those in the South-West that have experienced large immigration from the North-East. Thus, California, Arizona, Colorado and New Mexico had surfaces which were very different from the objective surfaces we are more familiar with. It seems that migrants from the North-East 'shrink' the distance to attractive locations. The same was true of migrants from New York to Florida. States having the least warped surfaces, on the other hand, were North Carolina, Tennessee, Kentucky and Virginia. It would be interesting to examine changes in such surfaces through time and also to construct such surfaces within urban areas for other types of spatial flow besides migration.

## 5.5   Tobler's Derivation of Migration Attractivities

In Section 3.2.2 and Table 3.4 we described a method of estimating the unknown attraction of a destination from a production-constrained interaction model. Tobler (1979) proposed an alternative method based on net migration rates. Define the net migration between i and j, $N_{ij}$, as

$$N_{ij} = T_{ij} - T_{ji}. \tag{5.12}$$

Tobler proposes that an estimate of this measure can be obtained from:

$$N_{ij}' = s(A_j - A_i)/d_{ij} \tag{5.13}$$

where A represents an attractiveness measure and s is a scaling constant determined by the units in which separation is measured. It can be ignored in the estimation of the As which are determined to a multiplicative constant. In fact, Tobler's formula is better represented as

$$N_{ij} = s(A_j - A_i)/d_{ij}^{\beta} \tag{5.14}$$

which generalises the distance-decay parameter. The overall net migration at place i is given by

$$N_i = \sum_j N_{ij} \tag{5.15}$$

which by substituting (5.14) yields

$$N_i = \sum_{\substack{j \\ j \neq i}} (A_j - A_i)/d_{ij}^{\beta}. \tag{5.16}$$

On transforming this equation, an estimate of $A_i$ is

$$A_i = \frac{\sum_{\substack{j \\ j \neq i}} A_j/d_{ij}^{\beta} - N_i}{\sum_{\substack{j \\ j \neq i}} d_{ij}^{-\beta}} \qquad (i=1,...,m) \tag{5.17}$$

By setting $A_1=0$, the set of $A_i$s can be solved since there are m-1 unknowns and m-1 equations. As Tobler (1979) notes, this equation allows the estimation of attractivity without knowing the individual flows - only the net changes for each origin, the $N_i$s, are needed.

Tobler provides an example of the application of equation (5.17) to 1965-70 US interstate migration between the 48 contiguous states. Rather than obtaining an estimate of the distance-decay parameter, he produced attractivities for two values of β (β=1 and β=2). These attractivity values, normalised to lie between 0 and 100, are presented in Table 5.2. In both cases, the attractivity of New York was set to 0 so that all the other values are relative to this. It can be seen that the two parameter values, although yielding different values for almost every state do produce much the same overall results (the Pearson correlation coefficient for the two sets was 0.976). In both cases, the least attractive states for US migrants during this period were clearly those in the North-East and Mid-West while the most attractive were Florida and those in the South-West.

**Table 5.2 Tobler's Estimates of US Attractivities**

| State | ß = 1 | ß = 2 | State | ß = 1 | ß = 2 |
|---|---|---|---|---|---|
| Ala. | 28 | 37 | Neb. | 30 | 39 |
| Ariz. | 52 | 56 | Nev. | 44 | 48 |
| Ark. | 32 | 42 | NH | 30 | 43 |
| Calif. | 92 | 86 | NJ | 29 | 43 |
| Col. | 43 | 51 | N.Mex. | 30 | 39 |
| Conn. | 29 | 42 | NY | 0 | 0 |
| Del. | 30 | 42 | NC | 34 | 45 |
| Fla. | 100 | 100 | ND | 28 | 38 |
| Ga. | 39 | 49 | Ohio | 25 | 36 |
| Idaho | 37 | 43 | Okla. | 34 | 43 |
| Ill. | 19 | 27 | Ore. | 49 | 52 |
| Ind. | 30 | 39 | Pa. | 19 | 27 |
| Iowa | 27 | 37 | RI | 29 | 41 |
| Kan. | 30 | 39 | SC | 34 | 44 |
| Ky. | 29 | 39 | S.Dak. | 29 | 39 |
| La. | 31 | 40 | Tenn. | 32 | 43 |
| Me. | 28 | 40 | Tex. | 49 | 56 |
| Md. | 32 | 46 | Utah | 36 | 43 |
| Mass. | 27 | 38 | Vt. | 30 | 42 |
| Mich. | 30 | 42 | Va. | 34 | 46 |
| Minn. | 31 | 41 | Wash. | 65 | 65 |
| Miss. | 30 | 39 | W.Va. | 26 | 36 |
| Mo. | 32 | 42 | Wis. | 30 | 41 |
| Mont. | 32 | 40 | Wyo. | 34 | 42 |

5.6   An Analysis of Migration in England and Wales

The analysis of migration within England and Wales which is now described is undertaken for two reasons. One is to demonstrate the information on destination attributes which affect a migrant's spatial choice that can obtained through the calibration of spatial interaction models. The other is to examine the hypothesis that individuals make spatial choices in a hierarchical manner as discussed in Chapter 4.

The following reports the results of calibrating several spatial interaction models using data on migration flows between 30 large labor market areas in England and Wales between 1980 and 1981. These labor market areas, termed Functional Regions (FRs), were devised by the Centre for Urban and Regional Studies at the University of Newcastle upon Tyne (Coombes et al., 1982). The 280 Functional Regions form an exhaustive and non-overlapping set of regions within the United Kingdom and each is sufficiently large to ensure that reported flows between the regions capture a large proportion of interurban flows while excluding most intraurban flows. A (30 x 29) matrix of flows between the 30 largest FRs was analysed using various data on these regions reported in Champion et al., 1987). The flows were first divided by 2.68, the average household size in 1980-81 (General Household Survey, 1982), in order to ensure that we were modelling the correct number of spatial choices. Clearly, in migration, family units move and the flows reported initially were in individual counts rather than family units. While this adjustment does not affect the resulting parameter estimates, it produces more accurate standard errors of the parameter estimates if the model is calibrated by Poisson regression (see Davies and Guy, 1987, for two alternative correction procedures). The spatial distribution of the 30 FRs used in the analysis is shown in Figure 5.3.

The variables used to explain migrants' destination choices in this system are listed in Table 5.3 and for the most part are self-explanatory. Spatial separation is defined by the straight-line distance between the centroids of the FRs. In migration studies, spatial separation influences destination choice through reducing information, and thereby increasing uncertainty, about places. It is not meant to be a surrogate for the cost of moving from one place to another. The centrality (destination competition) variable is included in the model framework in order to examine the extent to which destination choices are made hierarchically (see Chapter 4). It is expected that its associated parameter estimate will be significantly different from zero indicating that choice results in part from a hierarchical process. Notice that the centrality of a destination j is calculated with respect to all the other alternative destinations and not just to those included within the sampling framework described above (that is, the 30 FRs). The contiguity measure, a dummy variable being 1 if two FRs are contiguous and 0 otherwise, captures the effect of short distance migration across the boundary between two adjacent FRs and which may not be a result of the same type of spatial choice process as is examined in the rest of the flow matrix.

Figure 5.3  **The 30 Largest Functional Regions in England and Wales**

Migration between contiguous FRs could result merely from a change of residence rather than a change of employment. The inclusion of some of the remaining site variables (the $w_j$s) in the model is rather speculative. It is interesting to discover to what extent these various economic and social conditions affect destination choice.

The interaction model calibrated with these data is:

$$T_{ij} = \frac{\sum_{h=1}^{8} \exp(\alpha_h w_{jh}).\exp(\sigma b_{ij}).d_{ij}^{\beta}.c_{ij}^{\theta}}{\sum_{k=1}^{30}\sum_{h=1}^{8} \exp(\alpha_h w_{kh}).\exp(\sigma b_{ik}).d_{ik}^{\beta}.c_{ik}^{\theta}} \qquad (5.18)$$

where $T_{ij}$ represents the number of households in i choosing j; $w_{jh}$ represents a site attribute of destination j; $d_{ij}$ is distance; and $c_{ij}$ is the centrality of j with respect to all other possible destinations. Three of the site variables, population, change in unemployment rate, and mean house price, were entered in logarithms so that the parameter estimates for these variables are elasticities and are comparable across different systems. The remaining site variables were entered in a linear form since they were proportions and hence had a restricted range so that the exponential parameters estimated for these variables would automatically be comparable across systems. The model was calibrated by Poisson regression with the computer package GLIM (Payne, 1985); a method equivalent to maximum-likelihood estimation (Baxter, 1984).

Table 5.3  **Attributes of Functional Regions Used to Explain Destination Choice**

| | | |
|---|---|---|
| 1. $p_j{}^*$ | - | 1981 Population |
| 2. $d_{ij}{}^*$ | - | Straight Line distance between FR centroids |
| 3. $c_{ij}{}^*$ | - | Centrality Index $c_{ij} = \sum\limits_{\substack{k=1 \\ k \neq j \\ k \neq i}}^{280} p_k/d_{jk}$ |
| 4. $b_{ij}$ | - | Contiguity Measure (1 if i and j are contiguous; 0 otherwise) |
| 5. $o_j$ | - | Proportion of population classed as elderly ($>65$) |
| 6. $e_j$ | - | Proportion of population classed as economically active |
| 7. $m_j$ | - | Proportional change in employment 1978-81 (performance indicator) |
| 8. $u_j{}^*$ | - | Change in unemployment rate 1971-81 (performance indicator) |
| 9. $t_j$ | - | Proportion of population in managerial and professional socio-economic groups |
| 10. $a_j$ | - | Proportion of households with 2 or more cars |
| 11. $h_j{}^*$ | - | Mean house price |

*Indicates a variable whose functional form in the Poisson Regression Model is a natural logarithm

The parameter estimates obtained in the calibration of equation (5.18) with the 30 x 29 matrix of migration flows are presented in Table 5.4 along with their standard errors and significance. The interpretation of these results is that, ceteris paribus, migrants appear to be more likely to select regions that have large populations (and presumably that have large numbers of opportunities for employment, social activities, etc.); that are in close proximity; that are relatively isolated; and that are relatively prosperous. The only two variables that do not appear to determine migrants' choices are the proportion of the elderly and the proportion of higher socioeconomic group in a destination. The significant negative parameter on the centrality variable supports the hypothesis presented in Chapter 4 that spatial choice results from a hierarchical process. Individuals appear to make a destination selection from a restricted choice set (a spatial cluster of alternatives) and tend to underestimate the numbers of opportunities in large clusters. Support for this hypothesis can also be gained from further analysis of the data which is now described.

**Table 5.4  Parameter Estimates Obtained from the Full Migration Data Set**

| VARIABLE | PARAMETER | SE | SIG. |
|---|---|---|---|
| $p_j^*$ | 0.8853 | .0118 | + |
| $d_{ij}^*$ | -0.7875 | .0075 | + |
| $c_{ij}^*$ | -0.8855 | .0324 | + |
| bij* | 0.8217 | .0182 | + |
| $o_j$ | -0.2098 | .3575 | |
| $e_j$ | 0.4393 | .1655 | + |
| $m_j$ | 1.4460 | .1843 | + |
| $u_j^*$ | -0.3639 | .0303 | + |
| $t_j$ | 0.6023 | .5168 | |
| $a_j$ | 1.1260 | .3017 | + |
| $h_j^*$ | 0.2239 | .0701 | + |

Notes:

*   Indicates a variable whose functional form in the Poisson regression is a natural logarithm.

+   Indicates a variable whose parameter estimates is significantly different from zero at the 95% confidence interval.

Instead of calibrating the model in equation (5.18) on all the migration data, it can be calibrated separately with the choice data of migrants from each of the thirty origins. In this way, much more detailed information is obtained on the attributes of alternatives affecting migrants' spatial choices. The signs of the significant parameter estimates are given in Table 5.5 for each origin. What is interesting here is that only three variables, population, distance and centrality, appear to have any consistent influence on migrants' destination choice. All the other variables have very limited influence: that is, their parameter estimates are significant from some of the origins but not for others and even where they are significant, there are often different signs for different origins. Both the large number of occurrences of non-significance and the switching of signs are indications of variables whose influence on destination choice is relatively weak and in many cases may be spurious. The ability of the calibration to detect weak and possibly spurious relationships in this way supports the importance of the centrality variable which is significantly negative for all origins; again a result consistent with the hypothesis that individuals' spatial choices result from a hierarchical process.

### Table 5.5  The Significance of Origin-Specific Parameter Estimates

| FR | $p_j^*$ | $d_{ij}^*$ | $c_{ij}^*$ | $b_{ij}$ | $o_j$ | $e_j$ | $m_j$ | $u_j^*$ | $t_j$ | $a_j$ | $h_j^*$ |
|---|---|---|---|---|---|---|---|---|---|---|---|
| PLYM | + | - | - | n.a. | - | + | + |  |  |  | - |
| NORW | + | - | - | n.a. |  |  |  |  |  |  |  |
| NEWC | + | - | - | n.a. | - |  | - |  |  |  |  |
| BOUR | + | - | - | + | - | + | - | + |  |  |  |
| HULL | + | - | - | n.a. | - | + |  | - |  |  |  |
| CARD | + | - | - | n.a. | - |  |  |  | + |  | + |
| LOND | + | - | - | + | + | - | - | + | + |  | + |
| MIDD | + | - | - | n.a. | - | + |  |  |  |  |  |
| BRIS | + | - | - | n.a. | - | + |  |  |  |  | + |
| SOUT | + | - | - | + | - | + | - |  |  |  |  |
| BRIG | + | - | - | n.a. |  |  |  |  |  |  |  |
| PORT | + | - | - | + | - |  | + |  |  |  | + |
| SEND | + | - | - |  |  |  | - | - |  |  |  |
| LEED | + | - | - | + |  |  |  | + | + | + | - |
| LEIC | + | - | - | n.a. |  |  |  |  | + |  |  |
| OXFO | + | - | - | n.a. |  |  |  |  | + |  | - |
| COVE | + | - | - | + |  |  |  |  | + |  |  |
| STOK | + | - | - | n.a. |  |  |  |  | + |  | + |
| NOTT | + | - | - | + |  |  | + |  | + |  |  |
| LIVE | + | - | - |  |  |  |  |  | + |  |  |
| WOLV | + | - | - |  |  |  | - |  |  |  | + |
| DERB | + | - | - | + |  |  |  |  |  |  |  |
| SHEF | + | - | - | n.a. |  |  | + |  | + |  |  |
| BIRM | + | - | - |  | + |  |  |  |  | + |  |
| READ | + | - | - |  | + |  |  |  |  | + |  |
| BRAD | + | - | - | + |  |  |  | + | + |  |  |
| MANC | + | - | - | n.a. |  |  | + |  |  |  | + |
| ALDE | + | - | - | + |  |  | + |  |  |  |  |
| LUTO | + | - | - | n.a. |  |  |  |  | + |  | - |
| BIRK | + | - | - |  |  |  |  |  |  |  | + |

\*   Indicates a variable whose functional form in the Poisson regression model is a natural logarithm. The other variables are proportions.

n.a. Variable not applicable (no contiguous region).

The criterion for significance is that the probability of making a Type I Error is less than 0.05.

Interestingly, the centrality parameter estimates exhibit a marked spatial trend. As shown by Figure 5.4, there is a significant tendency for the centrality parameter to become more negative as the centrality of the origin increases. Since increasingly non-zero values of the parameter indicate greater intensities of hierarchical choice (see Chapter 4), it appears that the residents of central locations process spatial information hierarchically to a greater extent than do residents of peripheral locations. Elsewhere (Fotheringham, 1988c) an explanation has been provided for this finding in terms of the relationship between the extent to which one processes spatial information hierarchically and the amount of spatial information one has. Processing information hierarchically is an efficient means of reaching a decision since it reduces the number of evaluations one has to make (Fotheringham, 1988c). Because information on a location tends to diminish as the distance from the location increases, it follows that the residents of central locations will have more spatial information than those of peripheral locations. Hence, residents of central locations will need to process their information hierarchically to a greater extent than do residents of peripheral locations.

**Figure 5.4  The Spatial Pattern of Origin-Specific Centrality Parameter Estimates**

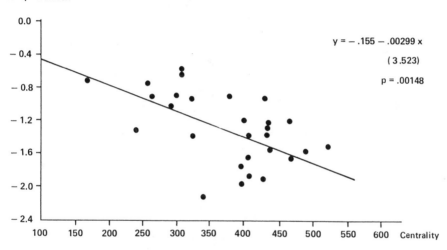

Centrality Parameter

$y = - .155 - .00299 x$

( 3 .523)

$p = .00148$

The extent to which misspecification bias is present in the parameter estimates of a spatial choice model not containing a destination centrality (or competition) variable can be seen in Figures 5.5a and 5.5b where the estimated origin-specific distance-decay parameters from two models are graphed against the location of each origin. In Figure 5.5a the estimated parameters are obtained from the calibration of a model without a centrality variable (but with all the other variables described in Table 5.3); in Figure 5.5b the parameter estimates are obtained from the same model to which the centrality variable has been added. The results in Figure 5.5a exhibit a trend (although only significant at the 90% confidence interval) found elsewhere (Fotheringham, 1981): the residents of more central locations appear to be less deterred by distance in their choice of destination. However, as demonstrated in Chapter 4, this is likely to be a spurious result caused by model misspecification. The more accurate parameter estimates, derived from the competing destinations model, confirm this by not exhibiting any spatial trend.

**Figure 5.5  The Spatial Pattern of Origin-Specific Distance Decay Parameter Estimates**

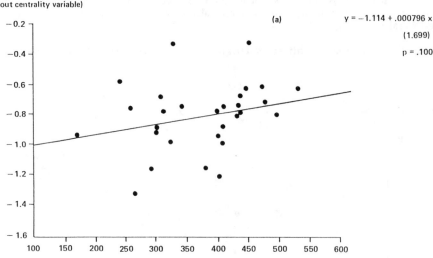

Distance — Decay Parameters
(without centrality variable)

(a)

y = −1.114 + .000796 x

(1.699)

p = .100

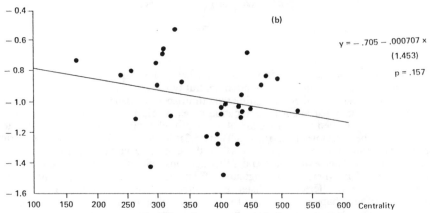

Distance — Decay Parameter
(full model)

(b)

y = − .705 − .000707 x

(1.453)

p = .157

Centrality

Table 5.6   **A Comparison of Two Sets of Models**

---

SET A:   Without Centrality Variable

| | | | | | | | | | |
|---|---|---|---|---|---|---|---|---|---|
| 1a. | $p_j^*$ | $d_{ij}^*$ | | | | | | | |
| 2a. | $p_j^*$ | $d_{ij}^*$(OS) | | | | | | | |
| 3a. | $p_j^*$ | $d_{ij}^*$ | $b_{ij}$ | $o_j$ | $e_j$ | $m_j$ | $u_j^*$ | $t_j$ | $a_j$ | $h_j^*$ |
| 4a. | $p_j^*$ | $d_{ij}^*$(OS) $b_{ij}$ | $o_j$ | $e_j$ | $m_j$ | $u_j^*$ | $t_j$ | $a_j$ | $h_j^*$ |
| 5a. | $p_j^*$ | $d_{ij}^*$ | $b_{ij}$ | $o_j$ | $e_j$ | $m_j$ | $u_j^*$ | $t_j$ | $a_j$ | $h_j^*$(OS) |

SET B:   With Centrality Variable

| | | | | | | | | | |
|---|---|---|---|---|---|---|---|---|---|
| 1b. | $p_j^*$ | $d_{ij}^*$ | $c_{ij}^*$ | | | | | | |
| 2b. | $p_j^*$ | $d_{ij}^*$ | $c_{ij}^*$(OS) | | | | | | |
| 3b. | $p_j^*$ | $d_{ij}^*$ | $c_{ij}^*$ | $b_{ij}$ | $o_j$ | $e_j$ | $m_j$ | $u_j^*$ | $t_j$ | $a_j$ | $h_j^*$ |
| 4b. | $p_j^*$ | $d_{ij}^*$ | $c_{ij}^*$(OS) $b_{ij}$ | $o_j$ | $e_j$ | $m_j$ | $u_j^*$ | $t_j$ | $a_j$ | $h_j^*$ |
| 5b. | $p_j^*$ | $d_{ij}^*$ | $c_{ij}^*$ | $b_{ij}$ | $o_j$ | $e_j$ | $m_j$ | $u_j^*$ | $t_j$ | $a_j$ | $h_j^*$(OS) |

---

% REDUCTION IN DEVIANCE WHEN $c_{ij}$ ADDED

| MODEL | 1 | 2 | 3 | 4 | 5 |
|---|---|---|---|---|---|
| % Reduction | 31.1 | 40.3 | 11.5 | 29.5 | 30.9 |

---

Note:

OS = Origin-specifice parameter estimates are obtained for all variables to the left of the bracketed OS.

* = A variable whose functional form in the Poisson regression model is a natural logarithm.

Finally, to demonstrate the importance of the destination centrality in determining spatial choice, the variable was added to five models which are described in Table 5.6. Model 1a is an extremely simple model containing only population and distance as explanatory variables (model 1b also contains the destination competition variable, $c_{ij}$). Model 2 is the same as model 1 but is calibrated separately for each origin so that origin-specific (OS) parameters are obtained. Model 3 contains all the explanatory variables listed in Table 5.3 but only one parameter is estimated for each variable. Model 4 is similar to model 3 except that origin-specific parameters are estimated for the population and distance variables. Model 5 is the full model calibrated separately for each origin. The increase in model accuracy when the destination competition variable is added to each model is given in Table 5.6 as the percentage reduction in deviance defined as

$$\%RiD = [(D_a - D_b)/D_a]*100 \qquad (5.19)$$

where $D_a$ is the deviance of a model which does not contain a competing destinations variable and $D_b$ is the deviance of the model with the addition of this variable. The deviance, a statistic given as standard output in GLIM and defined by Nelder and Wedderburn (1972) as:

$$D = \sum_i \sum_j T_{ij} \ln(T_{ij}/T_{ij}'), \tag{5.20}$$

is a measure of the accuracy of a model in replicating a known data set. Low values indicate an accurate model since D=0 when the model is perfect. Hence the large percentage reductions in deviance that occur due to adding the destination competition variable indicate the importance of this variable. The significance of the difference in the deviance measures can be examined with a chi-square test (see Nelder and Wedderburn, 1972) although this tends to be a relatively insensitive test. In all five cases, the difference in the deviance is significant.

## 5.7   An Analysis of Migration in the Netherlands

To reinforce the above results, a similar, although much simpler, analysis was undertaken using 1982 migration flows between the 51 housing market areas of the Netherlands (see Figure 5.6 and Table 5.7 for a definition of these areas). These data were supplied by the Central Bureau of Statistics (CBS) of the Netherlands. Each time an individual moves from one municipality to another in The Netherlands, he/she is required to complete a record of the move and these records are compiled by the CBS. The bureaucratic process in the Netherlands is such that these records are extremely accurate: individuals' addresses are checked on completion of other government documentation. Unfortunately, the only explanatory variables available for the set of housing market areas were population, a matrix of inter-area distances and destination competition (computed as in the analysis of UK migration). However, as the analysis of the UK migration in the previous section suggested, these may well be the only consistently relevant explanatory variables of the migration pattern. The spatial distribution of destination competition is presented in Figure 5.7 where it can be seen that the highest degree of competition is faced by locations in the central core of the country known as the Randstad and competition decreases fairly uniformly as distance from the Randstad increases.

Figure 5.6  **Dutch Housing Market Areas (see also Table 5.7)**

The two interaction models to be compared, one employing only population and distance; the other employing population, distance and destination centrality, are both production-constrained versions and are calibrated in this instance by maximum-likelihood estimation using the SIMODEL algorithm (Williams and Fotheringham, 1984) described in Chapter 3. Both models are calibrated separately for each of the 51 origins in the system so that two sets of origin-specific parameter estimates are obtained. The destination competition variable, measured as the accessibility of a destination to all its competitors, is again significantly negative for the large majority of origins (46 out of the 51) indicating the prevalence of hierarchical choice and competition effects. The spatial pattern of these parameter estimates (Figure 5.8) is identical to that found in the UK migration analysis (Figure 5.4) with parameter estimates being

## Table 5.7  Dutch Housing Market Areas

| | |
|---|---|
| 1. N. Groningen | 26. Amsterdam |
| 2. O. Groningen | 27. Waterland |
| 3. Leeuwarden | 28. Hoorn |
| 4. PR Wadden | 29. Kop van N. Holland |
| 5. ZW Friesland | 30. Alkmaar |
| 6. ZO Friesland | 31. Haarlem |
| 7. N. Drenthe | 32. Leiden |
| 8. Assen | 33. Den Haag |
| 9. Emmen | 34. Gouda |
| 10. ZW Drenthe | 35. Rijnmond |
| 11. NW Overijssel | 36. Alblasserwaard |
| 12. Zwolle | 37. Dordrecht |
| 13. Twente | 38. Goeree Overflakkee |
| 14. M. IJssel | 39. N. Zeeland |
| 15. O. Veluwe | 40. M. Zeeland |
| 16. Achterhoek | 41. Zeeuwsch Vlaanderen |
| 17. Arnhem | 42. Roosendaal |
| 18. W. Veluwe | 43. Biesbosch |
| 19. Nijmegen | 44. Breda |
| 20. Betuwe | 45. Tilburg |
| 21. O. Utrecht | 46. Den Bosch |
| 22. Utrecht | 47. Eindhoven |
| 23. W. Utrecht | 48. O. Brabant |
| 24. Flevoland | 49. Helmond |
| 25. Gooi | 50. N. Limburg |
| | 51. Z. Limburg |

increasingly negative as origin centrality increases. The trend of this relationship has a probability of less than .00001 of occurring by chance. Again then, there is strong evidence to suggest that the residents of centrally located origins, having relatively large amounts of spatial information, process this information hierarchically to a greater extent than do the residents of more peripheral regions.

The spatial pattern of the two sets of distance-decay parameter estimates (Figure 5.9) is also the same as identified in the UK analysis (Figure 5.5). When estimated in the calibration of the model without the destination competition variable, the parameter estimates become less negative as origin centrality increases (Figure 5.9a). This relationship is very significant (p=.00506). However, when the misspecification bias is removed by the addition of the destination competition variable to the model, this relationship disappears and there is no spatial pattern to the estimated distance decay parameters (Figure 5.9b).

Figure 5.7  **The Spatial Distribution of Destination Competition**

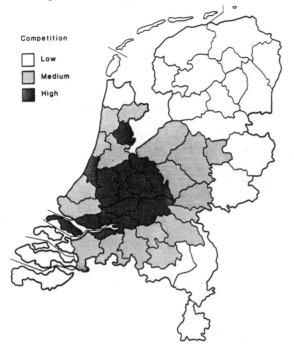

Figure 5.8  **The Spatial Pattern of Centrality Parameter Estimates**

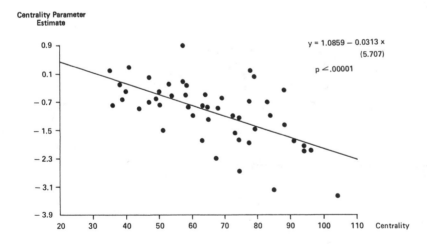

Figure 5.9  **The Spatial Pattern of Distance-Decay Parameter Estimates**

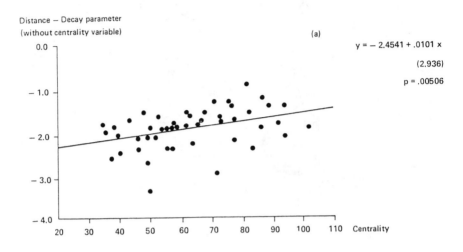

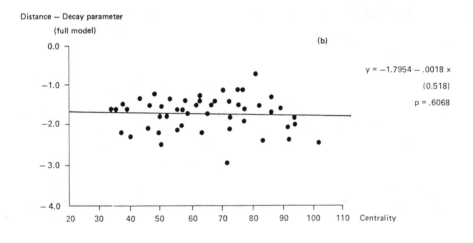

The spatial variation of the bias in distance-decay parameter estimates obtained from the simple gravity-type model is also shown in Figure 5.10 where a positive bias indicates an estimated distance-decay parameter which becomes less negative subsequent to the addition of the destination competition variable to the model. Clearly, the greatest misspecification bias is found for the most central origins. An explanation has been provided for this in Chapter 4 and elsewhere (Fotheringham, 1984).

Figure 5.10  **The Spatial Pattern of the Bias in Distance-Decay Parameter Estimates**

Bias in Gravity
Parameter Estimate

□ < 0

▨ 0.1 to 0.3

▧ > + 0.3

## 5.8 Summary

In this chapter we have discussed the importance of understanding the determinants of migration patters and the role that spatial interaction modelling can play in furthering this understanding. We illustrated this discussion with two empirical analyses: one on migration between Functional Regions in England and Wales; the other on migration between Housing Market Areas in The Netherlands. In both analyses the importance of including a destination competition variable was highlighted. The UK analysis, in particular, demonstrated that only three variables appear to be consistent determinants of the destination choices made by migrants: population, distance and destination competition. The first two variables are standard components of most spatial interaction models; the rationale for the third is given in Chapter 4 in terms of measuring the extent to which individuals' spatial choices result from hierarchical spatial information processing.

# CHAPTER 6

---

## APPLICATIONS TO RETAILING I: SINGLE PURPOSE TRAVEL

### 6.1 Introduction

The application of spatial interaction models in marketing and retailing is centered around the ability of certain models to answer the basic questions of why and how consumers select certain stores. Consequently, the model most frequently encountered is the production-constrained or spatial choice formulation given in equations (2.21) and (4.7), respectively. Apart from answering the basic question of "who shops where?", this model can be used in the following ways:

(i) To determine the attributes of a store that are important in attracting consumers. To what extent, for example, can low prices offset the disadvantage of a poor location or a low variety of products? Alternatively, to what extent can price levels at a store be raised without a significant drop in patronage?

(ii) To examine the market characteristics of an existing or proposed store. A store choice model, for example, can be used to forecast the proportion of a store's potential customers who belong to a particular socio-economic group known to have different purchasing habits from the rest of the population.

(iii) To relate the market characteristics of a store to the sales of particular brands or types of product. For instance, consumer purchasing patterns of brands and especially products are likely to vary by income, age, etc. and a knowledge of a store's market characteristics would be useful at two different levels. It would be useful for individual stores to know their market breakdown as a guide to the quantity of products to carry and to help in the decision of whether or not to carry a new item. The sales potential of the new item can then be gauged from the store's market characteristics. At another level, it would be useful for companies supplying particular brands of goods to have knowledge of the market characteristics of stores so that relationships between sales of particular products and socio-economic characteristics can be accurately established.

(iv) One of the major uses of store choice modelling is in determining the optimal location of a new store. This optimal location may be defined in terms of a maximum potential share of the total number of consumers in an urban area or, in the case of specialised stores, the maximum potential share of a certain sub-group of the population. In either case, it is necessary to know the determinants of consumers' shopping behaviour in order to assess the potential attraction of a new location.

(v) To determine the effects on sales at existing stores of opening a new store or closing an existing store.

(vi) To determine the optimal size of a new store at a given location (see Haynes and Fotheringham, 1984, for an example) or, similarly, determining the minimum profitable store size at a given location.

(vii) To examine the effects of increasing store size on market share. For a company owning several stores in an urban area, this would be useful in order to examine in which store an investment in extra size would be most profitable.

(viii) To provide a means of assessing the performance of individual stores.   Actual turnover figures can be compared to those predicted by the model.

(ix) To analyze the factors leading to the closure of a store.

(x) To derive trading areas around stores. Apart from being useful information in itself, the calculation of accurate trade areas can aid in the performance of other marketing models such as the NBD-Dirichlet model of consumer purchase frequency (Wrigley and Dunn, 1984).

The importance of the above questions is borne out by the dynamism of the retailing sector. In the United States, for example, Winn Dixie, a major food retailer with stores in fourteen states in the south-eastern and south-western US, opened 52 new stores in a 28-week period during 1985 and planned to open 95-98 new stores while closing 50-55 during fiscal year 1986-87 (information obtained from a Merrill Lynch Investment News Report). On a slightly smaller scale, in the United Kingdom, the Argyll Group opened 17 food superstores in 1987 and closed 9 while Sainsbury's opened 15 food superstores in both 1985 and 1986 (information obtained from company reports).

The operation of a store choice/spatial interaction model in a retailing environment is now examined in order to demonstrate how some of the above questions can be answered. Following this, the chapter is concluded with a discussion of how a production-constrained spatial interaction model can be used to provide an explanation for very rapid changes in retailing systems.

## 6.2   The Store Choice Model

It has already been established (Chapters 2 and 4) that out of a group of $O_i$ individuals at location i, the number selecting store j, $T_{ij}$, can be represented as

$$T_{ij} = O_i[\textstyle\prod_k x_{jk}^{\alpha_k}].\exp(-\beta c_{ij})/\textstyle\sum_h[\textstyle\prod_k x_{hk}^{\alpha_k}].\exp(-\beta c_{ih}) \qquad (6.1)$$

where $x_{jk}$ represents the level of the kth attribute of store j and $c_{ij}$ represents the cost of overcoming the spatial separation between i and j. The parameters $\alpha_k$ and $\beta$ reflect the relationship between particular store attributes and the probability of a consumer selecting that store. The basis of the use of this model in retailing is that, once calibrated, estimates of patronage can easily be obtained by summing over all origins. That is,

$$P_j = \textstyle\sum_i T_{ij} \qquad (6.2)$$

where $P_j$ represents total patronage at store j.

It is a simple matter to disaggregate these values by person-type m, so that

$$T_{ij}^m = O_i^m[\textstyle\prod_k x_{jk}^{\alpha_{km}}].\exp(-\beta_m c_{ij})/\textstyle\sum_h[\textstyle\prod_k x_{hk}^{\alpha_{km}}].\exp(-\beta_m c_{ih}), \qquad (6.3)$$

$$P_j^m = \textstyle\sum_i T_{ij}^m , \qquad (6.4)$$

and,

$$P_j = \textstyle\sum_m P^m . \qquad (6.5)$$

In order to calibrate and apply this model, however, several operational considerations need to be addressed. In particular, what are the relevant store attributes (often referred to as store image variables) and the relevant socio-economic categories (identified by the notation m above) to include in the model? We now address these two issues separately.

## 6.3   Store Attributes and Store Choice

Studies of store attribute or image variables are fairly common (see Hudson, 1974; Hubbard, 1978; Lincoln and Samli, 1979). Generally, information on store image is obtained from administering questionnaires in which consumers are asked to rank the store attributes they perceive as important in their selection of stores. Louviere and Meyer (1979), for example, surveyed 100 residents in Tallahassee, Florida, and 100 residents in Laramie, Wyoming. Three attributes overwhelmed all others in frequency of response by consumers: variety of products, convenience to residence, and price levels. Similarly, Schuler (1979) surveyed consumers in Bloomington, Indiana, and found that the most important store attributes were price levels, quality of merchandise, proximity of parking, quickness of service and proximity to residence. Recker and Kostyniuk (1978), from 300 completed questionnaires in Buffalo, New York, found the important store attributes to be divided into four major categories; quality of store, accessibility, convenience and service. The quality of store variables included prices of goods, variety of goods, and quality of goods. The accessibility variables were distance from residence and distance from work (the former, presumably, for major purchases and the latter for minor purchases). The convenience variables included parking facilities, proximity to other shops, convenient hours, degree of crowding in the store and the display of goods. Finally, the service variables included the acceptance of credit cards, check-cashing policy and the ease of returning goods.

While questionnaire surveys regarding store attributes can be useful, the danger in relying on them is that consumers may not be fully aware of what affects their choice of stores or they may be unwilling to divulge certain determinants of their shopping behaviour. An example of the former is the variable measuring proximity of a store to its competitors which has been hypothesised by Fotheringham (1983a; 1983b; 1986) to play an important, although possibly subconscious, role in determining destination choice. This was also described in a retailing context by Fotheringham (1985). The proximity of a store to its competitors can be described by a weighted inverse distance measure where the weight is store size. An economic rationale for the inclusion of this variable is that consumers may want to minimise the distance they have to travel to an alternative store in the event demand is unsatisfied at their first choice.

An example of the latter danger of relying on questionnaire surveys concerns certain prejudices in store choice, which consumers are very unlikely to acknowledge. In the United States, for example, for racial reasons some whites may tend to be reluctant to shop at stores in predominantly black neighbourhoods, and blacks may tend to be reluctant in shopping at stores in predominantly white neighbourhoods (Lloyd and Jennings, 1978). Similar consumer behaviour is likely to occur if the population is segregated by some trait other than race, such as by religion or language.

The store attribute variables identified above fall into four categories:

(i)   those that are impossible to quantify, such as the ease of returning goods, the friendliness of store personnel and the display quality of goods;

(ii)   those that can be quantified but for which accurate data are extremely difficult to obtain, such as the degree of crowding and price levels;

(iii)   those that are highly correlated with another, more easily measurable, variable, such as the variety of goods and the number of cashiers, which tend to be highly correlated with store size;

(iv)  those that suffer from none of the above problems.

Generally the variables falling into the last category, which are those most frequently encountered in retail studies, are the spatial separation of the store from the consumer's residence, store size, proximity of the store to similar stores, and the racial environment of the store. The spatial separation of the store and the consumer's residence can be measured by distance, which is a surrogate for a consumer's travel time and for the rate of information decay. Consumers are likely to be more familiar with local stores (Hanson, 1977) and are more likely to patronise stores with which they are familiar. The store size variable is a surrogate for the variety of merchandise available, and also possibly for general price levels since prices tend to decrease with increasing store size. The rationale for the inclusion of the other two variables has been described above.

Next an effort is made to evaluate and classify the literature in terms of the usage of variables from categories (i)-(iv) above. Section 6.3.1 reviews examples of studies using primarily engineering variables or proxy variables, (i.e. categories (iii) and (iv)). Section 6.3.2 discusses studies using qualitative characteristics, (i.e. category (ii)). Section 6.3.3 covers studies attempting to incorporate abstract perceptual data, (i.e. category (i)). Finally section 6.3.4 raises the difficult issue of incorporating differential preferences and prejudices into destination choice models.

### 6.3.1  Engineering Variables or Proxy Variables

Interaction models with behaviourally relevant independent variables (e.g. Cadwallader, 1981; Timmermans and Veldhuisen, 1981) relate choice behaviour to such generic variables as distance, parking availability and selection (number of stores in shopping area). McCafferty, O'Kelly and Webber (1979) report results from such a logit modeling exercise using a travel survey in Hamilton, Ontario. Data on grocery trip destinations from the survey were analysed using disaggregate destination choice models. Several different forms of these models were tested, but only one or two are reported here. The disaggregate model is as follows: for each individual k, let $J(k)$ be the set of destinations from which that individual may choose and $X_{jk}$ a vector of characteristics of destination j as perceived by k; then the probability that individual k chooses destination j is

$$p_{jk} = \exp(g(X_{jk}))/\sum_h \exp(g(X_{hk})) \tag{6.6}$$

where g is a linear function.  In equation (6.6) the vectors $X_{jk}$ contain the following variables:

$x_{jk1}$ = the travel time in minutes (by the chosen mode) of person k to shopping center j;

$x_{jk2}$ = 1 if destination j is the nearest shopping center to k's residence, 0 otherwise;

$x_{jk3}$ = 1 if destination j was a mall with free off-street parking, 0 otherwise;

$x_{jk4}$ = 0, 1, or 2, depending on the number of major (chain) supermarkets in the destination j;

$x_{jk5}$ = income class of person k if j is the nearest shopping center to k, and this variable is zero otherwise.

Table 6.1 presents the estimated coefficients and t-statistics for the model in equation (6.6). All the coefficients estimated using the survey data are significantly different from zero. Travel time from a shopping center depresses probability of visiting that center, as would be expected. If two shopping centers are identical in all respects except that one is 1 minute nearer person k than is the other, then the probability of k going grocery shopping at the nearer is 1.27 times the probability of visiting the more distant center. Similarly, the probability of visiting a center for groceries is enhanced by that center being the nearest center: a center which is one minute nearer to person k than another center and which is the nearest alternative to person k is 3.5 times more likely to be visited than the other, even if they are identical in all respects except location. Grocery shopping is primarily a local activity. Shopping centers which are malls with free off-street parking are nearly 10 times more likely to be visited than equivalent shopping centers which are not malls. Equally, shopping centers having major chain supermarkets are more likely if one chain is present and 3.2 times more likely if two chain supermarkets are present. Finally, the results imply that higher income people are more likely to visit their nearest shopping center than are low income people - perhaps because time is more significant (and price is less of a consideration) to high income people than to low income people. Other evidence on the role of income, price and convenience interaction is presented in Crafton (1979).

---

**Table 6.1 Estimated Logit Model Coefficients - Grocery destination choice**

---

| | Coefficients | t-statistics |
|---|---|---|
| $x_{jk1}$ | −0.24 | 12.69 |
| $x_{jk2}$ | 1.04 | 3.22 |
| $x_{jk3}$ | 2.29 | 8.77 |
| $x_{jk4}$ | 0.59 | 5.01 |
| $x_{jk5}$ | 0.80 | 2.27 |

$x_{jk1}$ = the travel time in minutes (by the chosen mode) of person k to shopping center j; $x_{jk2}$ = 1 if destination j is the nearest shopping center to k's residence, 0 otherwise; $x_{jk3}$ = 1 if destination j was a mall with free off-street parking, 0 otherwise; $x_{jk4}$ = 0, 1, or 2, depending on the number of major (chain) supermarkets in the destination j; $x_{jk5}$ = income class of person k if j is the nearest shopping center to k, and this variable is zero otherwise.

Source: From Table 2 of unpublished manuscript based on McCafferty, O'Kelly and Webber (1979).

---

Some evidence as to the transferability/stability of such a disaggregate demand model is provided by the fact that equation (6.6) calibrated to the base data (June survey) was able to predict the grocery destination choices of a later survey group almost as well as it could replicate the base data. (An average error of prediction of 0.27 compared with 0.23 in the first survey.) A similar effort to calibrate a non-grocery shopping model was not so successful, and this experience motivated more realistic approaches to multipurpose shopping trips (see Chapter 7). In summary, a grocery shopping model based on easily evaluated aggregate destination characteristics, and using an "alternative specific variable" as the only means of segmenting the sample proved to be a relatively good model of the destination choice of a sample of shoppers in Hamilton.

### 6.3.2 Qualitative Measures of Price, Service and Quality of Goods

Next we move to a discussion of factors which are of great importance to actual shopping behaviour: variables such as price, quality, and service which are very difficult to quantify. The influence of price, quality and service is largely unaccounted for in applied interaction models. Some analysis has been done with theoretical models (e.g. Griffith 1982; Haining,

1985) considering demand as a function of price, but overall the literature on spatial interaction has been unable to integrate this important factor in destination choice. Those studies which do focus on price and quality take a variety of forms:

(a) Studies of price variations (and non-price incentives such as coupons, discounts, trading-stamps etc.) across geographical locations.

The major concerns in such research are threefold. (1) To identify the role of the pricing mechanism of retail firms as an important competitive device; (for example: Andrews and Brunner 1975). (2) To represent adaptive responses in behavioural models of the retail firm (e.g adaptation to unforseen sales slumps; Cyert and March 1963). (3) To study the extent to which prices vary systematically across locations; this has been discussed by inter alia Ambrose (1979), Parker (1974), Hay and Johnston (1979; 1980), O'Brien and Guy (1985).

(b) Studies of the effects of price variations on behaviour.

The studies of the effects of price variations on behaviour are somewhat more dispersed, but they show that price is an important aspect of the attraction/sales of retail centers. Cadwallader (1981) has provided some evidence. The study examined the behaviour patterns of a homogeneous population of 53 households in West Los Angeles. The questionnaire asked the family member responsible for grocery shopping to describe their travel amongst a choice set of supermarkets and shopping centers. The responses indicate the variables which are important to consumers in their choice of center -- the important supermarket characteristics are checkout service, range of goods, quality of goods and prices. On the other hand the variables important in shopping center choice are parking facilities, prices, hours of opening and range of goods. Note that the study is highly aggregated over the sample; furthermore individual coefficients are not given for the independent variables; neither is it clear how distance is involved; finally it is not clear that price exhibits any systematic effect over the centers at different locations. Perhaps one would expect price to be the most important characteristic for further centers. As a general point this research does not handle the problem of trips which visited more than one location.

Many questions remain unresolved in the integrated modelling of price-sensitive spatial demand. Three issues in particular need further attention: (1) How do choice of center, subjective valuation of time, price level and consumer income interact? [See Crafton, 1979.] (2) To what extent is a consumer's behaviour influenced by the availability of accurate price information? [Devine and Marion (1979) study the impacts of price data on consumer behaviour and on retail pricing.] (3) Finally, what is the value of price information and how does consumer confidence in the accuracy of price information influence shopping center choice? [Goldman and Johansson (1978) have also raised this issue; see also the discussion of Hay and Johnston's (1979) model in Chapter 7.9.] In assessing the research in this area one feels that there are many theoretically challenging questions which have yet to be answered: prices, and their role in the consumers' behaviour continue to pose some important research questions.

### 6.3.3  Models with Composite Variables

The major difficulty with interaction models including price, quality and service variables is the need to identify standard bundles of goods. While it is difficult enough to collect price data, it is almost impossible to measure service/quality differences across an entire study area. This difficulty has been elegantly solved in the literature in a series of papers which combine attitudinal data with logit models of choice (see Recker and Kostyniuk, 1978; and Koppelman and Hauser, 1978).

As an example of the approach we consider McCarthy's (1980) definition of a shopping area's attraction as a linear combination of its variety, reliability, intra-center accessibility, price level and opening hours. Two sub-samples in the Los Angeles area were selected. It was found that the Mission district sample's destination choice was significantly influenced by trip convenience, trip safety, and shopping area mobility. The negative exponential function of distance had a positive coefficient as expected -- implying that the probability of choice decreases at a diminishing rate as distance to a center increases, (McCarthy, 1980, 1275). The second sample from the Southwest Area exhibited behaviour which is much less diversified than that of the Mission sample. 90% of the sample in the SW sample used an automobile to their favourite center, which in 77% of the cases was the Seramonte Center; furthermore this is the nearest center and 61% of the sample trips were made to that Center - evidently not much dispersion in store choice in this sub-sample. The only significant generic variables are a negative exponential function of distance and shopping area attraction.

We now attempt to relate the empirical work to a theoretical framework originally proposed by Koppelman and Hauser (1978). The translation of perceptions into preferences, and hence into choice, is analyzed in Koppelman and Hauser (1978) as a set of 3 stages. However, McCarthy (1980) focuses attention on the dimensions of attitudes to each alternative. Although McCarthy (1980) and Koppelman and Hauser (1978) claim that the use of the 'factors' underlying the data makes for easy interpretation/prediction, it seems that it is difficult to predict how a change in the physical characteristics of the system would alter the individuals perceptions of them. Are the factors stable? The relationship between physical characteristics of the area and shoppers perceptions of them is not clearly defined, (Timmermans and Veldhuisen, 1981, 1492). The work of McCarthy (1980) and Koppelman and Hauser (1978) has been primarily useful in elucidating the linkage between consumer perceptions, preferences and their choice behaviour. The techniques used have been successful in reducing a complex array of observations into a series of underlying dimensions, and using these dimensions to explain both preferences and choices. It seems clear that any small scale study with access to consumer attitude data should use these data to derive attributes of the alternatives. The evidence suggests that certain abstract shopping attraction variables are as important as quality and price. In the absence of consumer attitude data and given the difficulty of measuring the components of attraction directly, we are forced back to the conventional 'engineering' measures of shopping center size. We are unable to characterise the destination in terms of shoppers perceptions and attitude. The major reason for this difficulty lies in taking a city-wide sample and the difficulty of assembling a complete list of alternatives from which to evoke consumers' responses. Of course the weights attaching to attitudinal and perceived characteristics could be recorded through the revealed preference technique in Koppelman and Hauser (1978) - but this would certainly be reversing the order of analysis from preference-choice to choice-preference. In summary, an important extension to the travel demand modeling methodology has been suggested by Koppelman and Hauser (1978) who showed that perception and attitudinal data provide important 'extra' information to explain destination preferences and destination choice.

### 6.3.4 Measuring Racial/Cultural Prejudice

An attempt is now made to provide a means of modelling the reduction in the probability that an individual of type x will select a store when that store is located in a neighbourhood populated by individuals of type y. Many examples of such behaviour occur: Catholics and Protestants patronising stores in different sectors of cities in Northern Ireland (Boal, 1969); French and English speakers having different shopping patterns in eastern Ontario (Ray, 1967); differences between old order Mennonite and "modern" shoppers (Murdie, 1965); and a certain reluctance on the part of some wealthier individuals to shop in what are considered to be poorer neighbourhoods. A frequently encountered example of these behavioural differences occurs along racial lines, particularly in the United States, where whites and blacks exhibit different patronage patterns for stores in neighbourhoods of similar racial composition (Lloyd and Jennings, 1978).

To model the effect on shopping patterns of any prejudice to this type, consider individuals classified as belonging to group x or group y. Different specifications of the spatial interaction shopping model must then be used for the two groups because one of the store attributes for individuals belonging to group y would be $f(p_j^y)$, a function of the proportion of group y individuals living within a certain radius around store j, whereas one of the attributes for individuals belonging to group x would be $f(p_j^x)$. Since the range of $p_j^x$ and $p_j^y$ is a constant, 0-1, f is most appropriately defined as an exponential function so that:

$$f(p_j^y) = \exp(\lambda_x p_j^y) \tag{6.7}$$

and

$$f(p_j^x) = \exp(\lambda_y p_j^x) \tag{6.8}$$

where $\lambda_x < 0$ and $\lambda_y < 0$ are parameters denoting the degree of prejudice by groups x and y, respectively. The more negative is $\lambda_x$, for example, the more individuals of type x are deterred from shopping at stores in neighbourhoods dominated by individuals of type y. Fotheringham (1988a) gives more detail in the use of this attribute of store choice.

## 6.4   Variation in the Parameters of a Shopping Model by Socio-Economic Type

Suppose the following four variables were selected for inclusion in a store choice model: the distance from a consumer's residence to a store; the size of a store; a variable depicting the racial composition of the zone in which the store is located; and a store competition variable. Each of these variables has an associated parameter which can vary by person-type, m, and hence the model could be calibrated separately for different socio-economic groups.

However, apart from the prejudice variable described above, there is relatively little empirical evidence regarding systematic variation across person-types in parameter values associated with the above variables. An exception to this is the distance parameter, which appears to vary systematically by income group. Generally, individuals with higher incomes appear to be less constrained by distance in making spatial choices than are people with lower incomes (Davies, 1969; Potter, 1977). This is probably because individuals with higher incomes tend to be more familiar with their urban environment (Horton and Reynolds, 1971; Smith, 1976) and they have greater rates of car ownership (Thomas, 1974). Consequently, in most applications of store choice models the only disaggregation by person-type that needs to be made is in terms of income and race/religion/language if this is thought to be appropriate.

## 6.5   An Example of Applications in Marketing

In order to describe more fully the various uses of the store choice model in retailing, an application, presented in more detail elsewhere (Fotheringham, 1988a), is now briefly described. The modelling framework described above was applied to supermarkets in Gainesville, Florida, to predict both the pattern of major grocery purchases in the city and the market characteristics of each supermarket. Consumers in this example were faced with a choice of thirteen major supermarkets within the city limits and three that lie just outside. The spatial distribution of supermarkets is shown in Figure 6.1, which also describes the location of a supermarket that recently closed (see Table 6.4), the locations of four possible sites for a new supermarket (see Table 6.3) and the thirty-nine neighbourhoods that form the origin zones in the study. Detailed socio-economic data are available for each of the neighbourhoods from the 1980 US Census. Data on the supermarkets were obtained from a grocery store survey firm.

**Figure 6.1  The Spatial Distribution of Supermarkets in Gainesville, Florida**

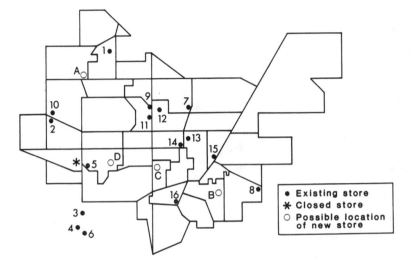

Five market characteristics, especially useful in estimating total sales and sales of individual products, were derived for each store. These were racial composition, age composition, one-person versus multi-person household composition, income composition, and renters versus owners composition. There tend to be differences in purchase frequency, for example, in snack food items between age groups; in convenience foods between one- and multi-person households; in specialty food items between income groups; and in pet foods between home-owners and renters. Knowledge of the proportions of a store's customers drawn from each of these groups would thus be useful to target the advertising of various products and also to determine sales potentials of foodstuffs at different locations proposed for a new store.

In order to give an idea of the information that can be obtained from the application of the store choice model in the system, the predicted market shares and market characteristics for three of the 16 supermarkets are presented in Table 6.2. On the basis of these figures, store 10 could be classified as serving predominantly white suburban home-owners with medium to high incomes; store 8 as serving predominantly black suburban families with low incomes; and store 5 as serving predominantly young renters with low incomes.

While such information is obviously useful to supermarket chains, the greatest utility of the modelling framework is when it is used for forecasting. To demonstrate this, Fotheringham examined four sites that seemed plausible locations for opening a new supermarket in Gainesville (see Figure 6.1). For each potential location, a full market analysis was carried out, the results of which are presented in Table 6.3. On the basis of potential market share, location C would be the recommended site for the new supermarket. However, other factors should also be considered. For instance, the market of a new store at location C would be dominated by low-income renters who may not spend as much money on grocery shopping as do other segments of the population.

**Table 6.2  A Market Comparison of Three Supermarkets**

| Market characteristic | Store | | |
|---|---|---|---|
| | 10 | 8 | 5 |
| Store size (000 ft$^2$) | 20.0 | 19.0 | 24.3 |
| Population served | 3,867 | 7,772 | 9,479 |
| Population served/1000 ft$^2$ | 193 | 409 | 390 |
| Market Share | 4.8 | 9.6 | 11.7 |
| % Black | 7.0 | 60.9 | 23.7 |
| % Age < 20 | 32.5 | 38.7 | 29.2 |
| % Age 20-9 | 22.8 | 24.5 | 42.3 |
| % Age 30-59 | 35.5 | 27.7 | 19.6 |
| % Age > 59 | 9.2 | 9.1 | 8.9 |
| % One-person households | 16.8 | 19.0 | 25.7 |
| % Low income[1] | 23.7 | 58.7 | 56.7 |
| % Medium income[2] | 50.4 | 34.8 | 33.1 |
| % High income[3] | 25.8 | 6.5 | 10.3 |
| % Renters | 31.0 | 50.5 | 70.2 |

Locations:
10 = Publix, NW 16th Blvd
8 = Winn Dixie, Hawthorne Road
5 = Winn Dixie, SW 16th Ave

[1] 1980 Annual family income < $15,000
[2] 1980 Annual family income $15,000-35,000
[3] 1980 Annual family income > $35,000

**Table 6.3  A Market Analysis of Four possible Locations for a New Supermarket in Gainesville**

| Market characteristic | Location | | | |
|---|---|---|---|---|
| | A | B | C | D |
| Population served | 3,716 | 4,952 | 5,670 | 4,829 |
| Market Share | 4.6 | 6.1 | 7.0 | 5.9 |
| % Black | 7.0 | 29.8 | 18.4 | 11.5 |
| % Age < 20 | 33.6 | 28.7 | 30.2 | 32.4 |
| % Age 20-9 | 23.9 | 32.6 | 42.7 | 32.4 |
| % Age 30-59 | 35.7 | 26.0 | 18.7 | 25.6 |
| % Age > 60 | 6.8 | 12.7 | 8.5 | 9.6 |
| % One-person households | 15.8 | 28.9 | 30.8 | 23.6 |
| % Low income | 25.0 | 51.8 | 52.5 | 36.9 |
| % Medium income | 58.6 | 38.8 | 35.7 | 40.9 |
| % High income | 16.4 | 9.4 | 11.8 | 22.2 |
| % Renters | 29.3 | 55.8 | 68.1 | 47.4 |

[1] Supermarket size in each case is 20,000 ft $^2$ (approximate average size of supermarkets in Gainesville)

One final application of market share modelling demonstrated here concerns a business failure. During the summer of 1985, a large supermarket owned by the Winn Dixie chain closed in Gainesville due to low profits (its location is shown in Figure 6.1). Fotheringham (1988a) analyzed why this business failed by estimating its market characteristics while in operation. This was done by adding this store to the data set described above and re-running the analysis. The market characteristics of the failed store (17) and three other Winn Dixie supermarkets that are successful are listed in Table 6.4. The estimated market share for store 17 is not particularly low (7.6 percent), but the store was large and the ratio of population served to unit area is not very high (229 per 1,000 square feet). On the basis of these figures, store 17, while not appearing to be a particularly profitable store, does not give any strong indication of being a business failure. However, the store differed from other Winn Dixie supermarkets in three critical market categories. Because it was located in the western part of the city, its market consisted of far fewer blacks, fewer low-income consumers and a greater percentage of high-income consumers than the successful Winn Dixie stores. These differences hint strongly at why store 17 failed. Winn Dixie supermarkets in Gainesville have, and tended to nurture due to their product lines, an image of serving low-income black consumers. A Winn Dixie store in a predominantly white, relatively wealthy area, such as store 17, would probably not attract as many customers as might be thought given no knowledge of the chain's local image. A detailed market analysis might have highlighted this point and might have prevented the locational decision that led to a business failure.

**Table 6.4  Predicted Market Characteristics of the Four Winn Dixie Supermarkets Prior to the Closure of Store 17**

|  | Store Number | | | |
|---|---|---|---|---|
| Market characteristic | 5 | 7 | 8 | 17 |
| Store size (000 ft$^2$) | 24.3 | 24.0 | 19.0 | 27.0 |
| Population served | 8,828 | 5,343 | 7,408 | 6,195 |
| Population served/1000 ft$^2$ | 363 | 223 | 389 | 229 |
| Market Share | 10.9 | 6.6 | 9.1 | 7.6 |
| % Black | 24.3 | 19.6 | 61.9 | 10.1* |
| % Age < 20 | 29.0 | 29.5 | 38.8 | 31.6 |
| % Age 20-9 | 42.4 | 31.4 | 24.3 | 33.6 |
| % Age 30-59 | 19.7 | 28.6 | 27.8 | 26.0 |
| % Age > 59 | 9.0 | 10.4 | 9.1 | 8.9 |
| % One-person households | 25.7 | 23.9 | 18.9 | 22.4 |
| % Low income | 57.0 | 42.4 | 58.9 | 35.1* |
| % Medium income | 33.0 | 46.9 | 34.7 | 44.5 |
| % High income | 10.1 | 10.6 | 6.4 | 20.5* |
| % Renters | 70.4 | 48.1 | 50.4 | 50.3 |

*Major differences

## 6.6  Deriving Potential Revenue Surfaces

In equations (6.2) and (6.5) it was noted that the total patronage at a new store could be predicted for any location in an urban area. Usually, as in the example given above, a limited number of specific locations are considered. For each of these a value of $P_j$ is then calculated as an aid to finding the most profitable location. Often, $P_j$ is converted to a monetary value by multiplying it by average shopping expenditure so that an expected revenue value is obtained. An alternative strategy, which has the same aim, is to produce a potential revenue surface that will indicate the revenue generated from locating an outlet at any location within an urban area. Such a surface is very easy to construct: a grid is laid over the urban area with known coordinates at every intersection on the grid. Each intersection is then treated as a possible location for the new outlet and a potential demand calculated at that point. This produces a matrix of potential demand values which can be either plotted as a surface (Fotheringham, 1985 provides several examples of such surfaces) or as an isoline map.

Fotheringham (1985) went a step further in the generation of revenue surfaces by producing surfaces not for the sales of an individual outlet but for the revenue share of a set of franchised stores. The scenario was as follows. Consider a city containing two groups of stores: one group belonging to company A; the other belonging to company B. Company A is planning to locate a new store in this city. Clearly, it is not interested only in the sales potential of the new store but also in how the new store will increase the overall market share of the company. It would clearly be an unprofitable venture to locate a store so that it drew heavily on customers already patronising stores owned by company A. Ideally, company A would like to locate the new store so that it draws heavily from company B's existing customers. Information on the drawing power of a set of stores can be obtained by calculating the projected sales at every store due to the location of a new store at location (x,y). Consequently, a surface can be produced depicting A's potential share of the total market obtained by locating the new store at any location within the urban area.

Figure 6.2, reproduced from Fotheringham (1985), represents three such surfaces under varying conditions. In (a), the location of existing stores does not affect the probability of a consumer choosing the new store: the surface is relatively flat. In (b) the probability of a consumer choosing the new store is increased if that store is located in close proximity to existing stores. As discussed above, such a situation results from consumers wanting to minimise their travel costs to alternative stores in the event that the one they originally selected cannot satisfy their demands. In (c), which is indicative of a competition effect, the probability of a consumer choosing the new store is decreased if that store is located in close proximity to existing stores.

## 6.7  Modelling Rapid Change in Retailing Systems

In the last thirty years, the locational pattern of retailing has undergone some dramatic changes. We have witnessed the replacement of the corner store by the supermarket, the clustering of outlets into shopping malls and retailing strips, and the reduction of downtown retailing dominance with the suburbanisation of retail outlets. These changes have not been continuous but have been discrete and therefore they cannot be modelled by conventional means. They have recently been investigated, however, using the theory of catastrophes and bifurcations. Harris and Wilson (1978), in a seminal article, explored explanations for discontinuities in retail size dynamics, while Fotheringham (1985) has discussed similar discontinuities in retail location dynamics. Here we will briefly review these developments.

## Figure 6.2   **Examples of Potential Revenue Surfaces**

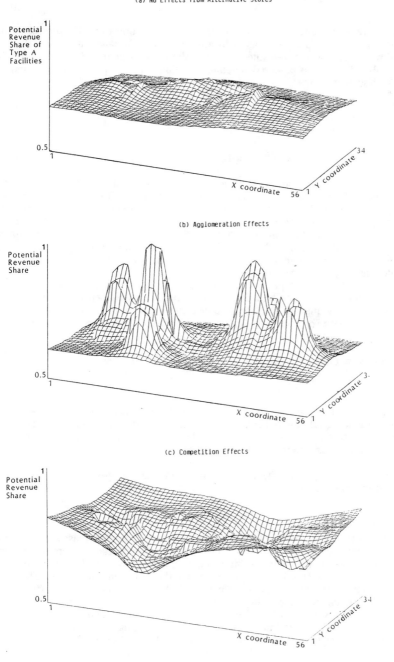

(a) No Effects from Alternative Stores

(b) Agglomeration Effects

(c) Competition Effects

The Harris-Wilson (H-W) model examines the behaviour of an equilibrium point between demand and supply connected with an individual retail outlet. Define the profit, $\pi$, associated with a retail outlet of type m at location j by

$$\pi_j^m = D_j^m - C_j^m \tag{6.9}$$

where $D_j^m$ is the revenue generated by a type m outlet at location j and $C_{jm}$ represents the operating costs incurred by a type m outlet locating at j. The demand side is modelled by summing the potential revenue generated by an outlet of a particular size at a particular location over all origins. The potential revenue is estimated by a production-constrained shopping model so that

$$D_j^m = \sum_i S_{ij}^m \tag{6.10}$$

where

$$S_{ij}^m = E_{im}P_i(w_j^m)^{\alpha_m}.\exp(-\beta_m c_{ij})/\sum_h(w_h^m)^{\alpha_m}.\exp(-\beta_m c_{ih}) \tag{6.11}$$

$S_{ijm}$ represents the amount of money spent at a type m retail outlet in location j by the residents of zone i; $E_{im}$ represents the average expenditure on retail goods of type m by a person living in zone i; $P_i$ represents the population of zone i; $w_{jm}$ represents the attractiveness of the retail outlet m at location j and is measured here by outlet size; $c_{ij}$ represents the cost of traveling from i to j; $\alpha_m$ is a parameter reflecting consumers' perceptions of the attractiveness of the size of type m retail outlets; and $\beta_m$ is a parameter reflecting consumers' perceptions of the disutility of overcoming the spatial separation of origins and destinations in order to purchase retail goods of type m. It is expected that $\alpha_m > 0$ and $\beta_m < 0$. The notation h represents a retail outlet so that the denominator is summed over all outlets including j. Henceforth, for simplification, it is assumed that only one type of retailing is being examined so that the m index can be dropped for convenience. This simplification avoids the problem of dealing with multipurpose trip-making which is discussed in detail in Chapter 7.

The supply side is modelled by assuming that $C_j$ is a function of $w_j$,

$$C_j = kw_j^\sigma \tag{6.12}$$

where $k > 0$ measures the operating costs per unit outlet size, here taken to represent the efficiency of operating retail outlets (as k decreases, the operation of retail outlets becomes increasingly efficient) and where $0 \le \sigma \le 1$ denotes the presence of internal scale economies (as $\sigma$ decreases, internal scale economies increase).

It is postulated that changes in the size of an outlet will be governed by the profit associated with that outlet,

$$\partial w_j/\partial t = f(\pi_j) \tag{6.13}$$

and that the size of the outlet increases when profits are positive and decreases when profits are negative. An equilibrium outlet size, $w_j^*$, is achieved when profits are zero. Harris and Wilson (1978) demonstrate that only one stable equilibrium exists when $0 < \alpha \le 1$ but two stable equilibria exist when $\alpha > 1$. What makes the equilibrium solutions of the differential equation in (6.13) interesting is that $\pi_j$ is a function of $D_j$ which, from equations (6.10) and (6.11), is a nonlinear function of $w_j$. The presence of this nonlinearity generates bifurcation properties in the system when $\alpha > 1$ and leads to the potential for discontinuous change in $w_j^*$ if it is assumed that the operation of retail outlets is becoming more efficient over time (i.e., k is decreasing over time). (See Harris and Wilson, 1978; Wilson, 1981a, 1981b; and

Fotheringham and Knudsen, 1984; 1986 for more details of this bifurcation point.) The value of $\alpha$ in equation (6.11) is thus critical in producing the necessary conditions to explain a discontinuous change in the equilibrium size of retail outlets. Only when $\alpha > 1$ can the discontinuous change shown in Figure 6.3 from a small to a large outlet take place. This change parallels what happens in reality when supermarkets replaces corner-grocery stores and when hypermarkets replaced supermarkets.

Fotheringham and Knudsen (1986) extended the H-W framework in three ways: (i) to allow for a relationship between consumption patterns and the agglomeration of retail outlets; (ii) to make the costs of operating a retail outlet be a function of locational rent and external scale economies; and (iii) to allow retail outlets to change location as well as size as a response to change in profit levels.

To incorporate extension (i), a variable was added to the demand model in equation (6.11) measuring the effect of competing destinations. As described above, this variable, $A_j$, denotes the accessibility or relative location of outlet $j$ to its potential competitors. It has an associated parameter $\theta$. As above, a positive value of $\theta$ indicates the presence of consumer agglomeration economies: the more accessible is outlet $j$ to other outlets, the greater the probability that individual $i$ will select outlet $j$, ceteris paribus. A negative value of $\theta$ indicates the presence of consumer competition forces between retail outlets: the more accessible is outlet $j$ to other outlets, the smaller the probability that individual $i$ will select $j$, ceteris paribus.

Figure 6.3  **Supply and Demand Equilibria when $\alpha > 1$**

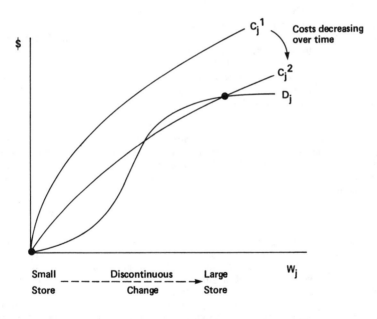

● = stable equilibrium point (see Harris and Wilson, 1978)

To incorporate extension (ii), the costs of operating a retail outlet are made not only a function of the size of the outlet but also a function of the location of the outlet with respect to its competitors. That is,

$$C_j = k_1 w_j^{\sigma_1} + k_2 A_j^{\sigma_2} \tag{6.14}$$

where $k_1 > 0$ denotes the efficiency of operating retail outlets; $\sigma_1$ denotes the presence of suppliers' internal scale economies; $k_2$ represents retail rent gradients around existing outlets (as $k_2$ decreases, it becomes relatively cheaper to locate in close proximity to existing retail outlets); and $\sigma_2$ denotes the presence of suppliers' external scale economies. As $\sigma_2$ decreases, suppliers' external scale economies increase.

Finally, to incorporate extension (iii), both locational and size dynamics are hypothesised to be functions of profit so that,

$$\partial w_j / \partial t = f(\pi_j) \tag{6.15}$$

and

$$\partial A_j / \partial t = f(\pi_j) \tag{6.16}$$

As an outlet's profits rise, the outlet can either increase its size and/or move to a more central location. Conversely, as an outlet's profits decrease it can either decrease its size and/or move to a more peripheral location. Equilibrium values of $w_j$ and $A_j$, $w_j^*$ and $A_j^*$, respectively, are achieved at zero profits.

Both the demand and supply surfaces are now functions of $w_j$ and $A_j$. The exact shapes of these surfaces depend on the parameters $\alpha$ and $\theta$ in the demand equation and on the parameters $k_1$, $k_2$, $\sigma_1$ and $\sigma_2$ in the supply equation. The parameters $\alpha$ and $\theta$ denote social conditions: $\alpha$ represents consumers' scale economies and $\theta$ represents consumers' agglomeration economies. The parameters $k_1$, $k_2$, $\sigma_1$ and $\sigma_2$ denote economic conditions as described above. The interaction between these parameters will lead to a variety of discontinuous changes in the equilibrium combinations of $w_j$ and $A_j$ when $\alpha > 1$ and $\theta > 1$ as described in detail by Fotheringham and Knudsen (1986).

Thus, by extending the Harris and Wilson model in this manner, it is possible to explain discontinuous change in either the size or the spatial distribution of retail outlets. With the latter, discontinuous change in relative location away from the origin represents change such as the rapid nucleation of outlets into malls; discontinuous change in relative location toward the origin represents change such as the decentralisation of retail outlets from downtown locations.

## 6.8   Conclusions

In this chapter we have examined the use of a choice-based spatial interaction model to provide information on several important retailing issues. We have considered the optimal location for a new store; the impacts of locating a new store on existing markets; the derivation of market shares for potential store sites; the derivation of potential revenue surfaces; and have provided an explanation for rapid changes in retailing systems. We now extend our discussion of the application of spatial interaction models to retail systems by examining the role multipurpose travel plays in determining shopping patterns and the implications this has for some of the issues described above.

# CHAPTER 7

## APPLICATIONS TO RETAILING II: MULTIPURPOSE TRAVEL

### 7.1 Introduction

In Chapter 6 existing spatial interaction models of the allocation of demand to retail centers in an urban area were discussed. The probabilistic allocation model described in Chapter 6 is flexible enough to allow for diversity in destination choices and so it improves over the strict assumptions built into alternative theories of consumer demand such as central place theory. It is also supported by evidence that consumers do travel further than the minimum necessary distance to purchase an item. For example, in a recent study in Gainesville Florida, Trew (1988) found that only 36.1% of shoppers patronized their nearest supermarket. (See also Scott, 1970, 15; Clark and Rushton, 1970; Fingleton, 1975; and Hubbard, 1978.) These observations provided the motivation for some theoretical work on consumer behaviour, (see for example Burnett, 1973; 1977 and 1978; and the choice theoretic discussion in Chapter 4). However, several complexities in the more general system of non-work interactions are not represented by those simple models. The present chapter elucidates these more complex interactions, and proposes a model to represent them.

While the probabilistic allocation model represents a major improvement over deterministic nearest center allocation, it is still an inadequate representation of some kinds of trip making because it is restricted to single-stop, single-purpose trips (see also Hanson, 1980a). The remainder of this section will show that intra-urban spatial interaction is a much more complex than the single-stop trip, and that multistop multipurpose trips are, in some cases, the predominant form of urban travel.

Hensher (1976) refers to several studies and concludes that at least 30% of all urban travel involves trips with more than two stops, and often involves more than one purpose (Hensher, 1976, 655). Hanson (1980a) summarises work which shows that in Lansing, Michigan 46% of all trips segments occurred during multipurpose travel and that in Uppsala, Sweden, as many as 61% of all movements were associated with multipurpose travel. Manheim (1979, 429) reports that a survey in Washington D.C., found that only 501 of 1259 shopping trips were of the HOME-SHOP-HOME type and the rest were part of multilink chains. Further evidence of multipurpose travel is provided in Adler and Ben-Akiva (1979). Thus, the frequency and importance of multipurpose trip making has been recognised for many years. In fact the early workers in the field of highway development and transportation planning were conscious of complex spatial interaction (see Marble, 1964; Nystuen, 1967, and Garrison et al., 1959, p. 223). More recently this theme has been expanded to include both empirical studies of daily activity patterns (Pas, 1982); theoretical examinations of spatial economic systems under relaxed rules of interaction (Mulligan, 1984; Eaton and Lipsey, 1982; Bacon, 1971; 1984); and, models of the location decision by firms under both competition and complementarity induced by comparison shopping (Eaton and Lipsey, 1976; McLafferty and Ghosh, 1986 and 1987). An excellent literature review surveying these contributions is in Thill and Thomas (1987).

These observations are important for every phase of the transportation planning process. Recall that transportation planning is often characterized as a set of interrelated stages: the generation of demand for travel; distribution of demand over destinations and modes; allocation of flows to specific routes and time-of-day categories; and, in some recent research, further modelling of the behaviour of the traveller up to the time of the next trip.

Perceptive analysts have noted that the occurrence of multistop multipurpose travel impinges on every step of this process (Hanson, 1979). Thus the trip generation decision is a complicated one involving a bundle of interrelated demands and constraints. Specifically, trips are formed in order to satisfy some needs (e.g. food shortage in Davies and Pickles, 1987) and must take place against the background of household transportation resources. When trips are designed to accomplish several tasks during the same travel pattern, some efficiency may be gained, and some inconvenience may be caused, by the difficulty of planning the entire trip. Trips may be dynamically rescheduled in order to take advantage of other unexpected (or postponed) opportunities. The trip generation decision should therefore be seen as intricately bound up in decisions about the efficient use of time, resources, and transportation facilities. Similarly, the trip distribution phase of the model is fundamentally concerned with a set of (one or more) destinations to be visited on the trip. As a trip progresses, accessibility to alternative locations increases, and the trip may be structured to take advantage of these contingencies. Finally, the route and time-of-day decisions are more complex in the presence of multiple destination travel patterns. Hanson (1979) has discussed a series of hypothetical travel behaviours which solve the path problem in a variety of ways.

The purpose of this chapter is to provide a survey of some the ideas and results which have been found in the literature on complex non-work spatial interaction. The chapter is in two main parts - sections 7.2, 7.3, and 7.4 comprise the first part and they are concerned with some empirical results from a travel diary survey in the Hamilton Ontario metropolitan area. This survey has been used extensively in analyses of non-work travel and interaction and so provides a good basis for several related case studies. The second part (sections 7.5 - 7.9) covers more theoretical issues and addresses implications for model design arising from observations about complex trip behaviour.

## 7.2   The Hamilton Travel Diary Survey

Some evidence from a travel diary survey from Hamilton, Ontario highlights the importance of complex spatial interaction, and in particular shows that multipurpose trip making is common. This section of the chapter describes the data used in the simple descriptive analyses of section 7.3 and 7.4. The study area is in the Hamilton Census Metropolitan Area (CMA) as defined by Statistics Canada in the 1976 Census, (see Figure 7.1). Hamilton CMA had an estimated population in 1978 of 534,300 with an estimated per capital income of $7,710.

The Hamilton-Wentworth area had a population of about 413,000 in 1978, residing in approximately 135,000 households. The main part of the survey was carried out in the highly urbanised section of the Regional Municipality of Hamilton-Wentworth centered on the City of Hamilton (1976 population, 312,005), Stoney Creek (1976 population, 30,290), Dundas (1976 population, 19,180) and Ancaster (1976 population 14,255). The remainder of the 409,490 people living in Hamilton-Wentworth in 1976 were in the predominantly rural sections of the region to the south and south-east of the City of Hamilton. The City of Burlington, in the Regional-Municipality of Halton is adjacent to the main study area and had a population of 104,315 in 1976. There is relatively little interaction between Hamilton and Burlington, due to the physical difficulty of travel between the areas. For these reasons, the City of Burlington is treated as a special external area, with which residents of the main study interact, but where the trip origins are ignored. The study area was divided into zones: zones 1-14 make up the City of Hamilton; the towns of Stoney Creek, Dundas and Ancaster are zones 15, 16 and 17, respectively, and the City of Burlington is divided into zones 18 and 19. Zone 20 represents external locations, (see Figure 7.2). (Zones 9, 10, 11, 12, 13 and 14 are known as the 'mountain', since they are separated from the rest of the city by the Niagara escarpment.)

The data were collected from a sample of 704 households who kept a travel diary for two weeks in the summer, 1978. The sample comprised of approximately 40 households in each of 17 Planning Divisions in the Hamilton Wentworth Region. The households recorded all travel made by their adult members and some summary measures of childrens' activities. The information pertaining to the adults' travel includes means of transport, purpose, destination, and expenditure for each stop of the trip. A summary of the survey and data collection method is given in Webber (1980).

Two general rules were followed in coding the diaries. The first concerns the definition of a separate stop. The actual address of the stop was reported, and any further activity was coded as being on the same stop if it took place within two blocks: otherwise it constituted a separate stop. This means that visits to several stores in the same shopping center are recorded as a single stop with several purposes, of which only the first purpose is considered here.

Figure 7.1 **Hamilton and Surrounding Communities**

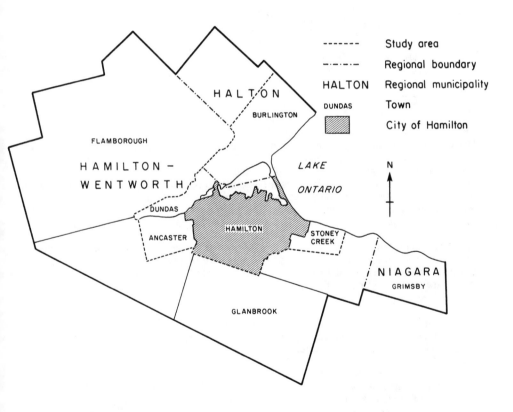

Figure 7.2 **Zoning Scheme for the Travel Diary Survey**

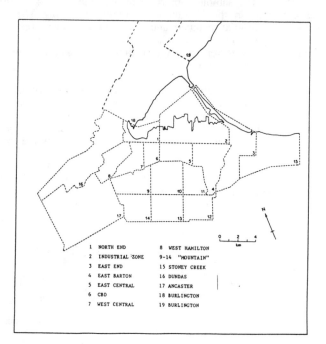

| | | |
|---|---|---|
| 1 | NORTH END | 8 WEST HAMILTON |
| 2 | INDUSTRIAL ZONE | 9-14 "MOUNTAIN" |
| 3 | EAST END | 15 STONEY CREEK |
| 4 | EAST BARTON | 16 DUNDAS |
| 5 | EAST CENTRAL | 17 ANCASTER |
| 6 | CBD | 18 BURLINGTON |
| 7 | WEST CENTRAL | 19 BURLINGTON |

The second deals with trips in which two persons set out together but then go their separate ways after some point.  For example, A and B set off together, A is dropped at the workplace, B goes on to another workplace and later both return home separately.  In this case, two trips are coded: (1) B (A is a passenger) goes to workplace (purpose: drop off A), goes to B's workplace (purpose: work) and returns home; (2) A goes to work (mode: passenger in B's car) and returns home.  In general, a separate trip is recorded for each adult resident who makes at least one stop independently of the other adults on the trip.

The results reported here are based on the 704 diaries collected from residents of zones 1-17.  Trips were considered only if they had their home base in zones 1-17.  Three other zones are treated as external zones, but it must be emphasised that the trips upon which the results are based could pass through external zones on any stop, provided the origin was home.

From these data, two simple components of multistop multipurpose trip making can now be illustrated: (1) the importance of multistop multipurpose trips as a percentage of all travel; and (2) the linkages between activities.

(1)  From the survey results 14,865 of all the stops occurred on single-stop single-purpose trips - that is, trip makers engaged in some activity and then returned home.  This is a large proportion of (65%) of all first stops, but only 41% of the total number of stops (36,103) made by the sample households.  This shows that although a high proportion of trip makers return home at any particular stop, the majority of trips involve more than one stop, thus 74% of all non-grocery shopping stops took place on other than a single-stop trip.  (Similar

results are reported in other studies using this data set e.g. Preston and Takahashi, 1983; Takahashi, 1986; and Williams, 1988.)  The exception is of course the work trip category, where the simple HOME-WORK-HOME structure accounted for 61% of all work stops. The implication of this simple analysis for existing urban non-work spatial interaction models is that a model of home-based single-stop trips would ignore over half the data, and would miss entirely the importance of the interconnection between activity locations. This simplification is probably a reasonable one for work trips and for many routine grocery trips such as discussed in Chapter 6, but is certainly not for comparison goods shopping trips. Additional discussion appears in Hanson (1980b) who considers the importance of multipurpose trips in the context of the intra-urban journey to work.

(2)  The second component, namely the linkages between different activities is illustrated in Tables 7.1-7.4.  These tables show the purpose at stop m classified by the purpose at stop (m - 1).  (Activity linkage data of this type have been used for many years; see Damm (1979) and Hanson (1979) for reviews.)  Generalisations from these tables are difficult since the connections between activities form a chain, and therefore the activity at any particular stop on that chain is conditioned upon previous purposes.  In general though, there is a large probability of interaction between the two types of shopping activities, and between social and recreational activity and other purposes.  There is also a large probability of continuing to further social-recreational activity following a stop for that purpose. Although work trips are predominantly single-stop, there is a noticeable connection between this and the social-recreational category.  Next, summary measures based on such activity linkage data are constructed following the ideas in Horton and Wagner (1969) and Hemmens (1970), who have studied similar data for Waco, Texas and Buffalo, New York.  Later, an improved model will be constructed which allows for a spatial component in interaction probabilities.

Given the evidence of the importance of multistop, multipurpose trips for urban spatial flows, the aim of the next section is to provide a simple accounting framework for the pattern of interaction levels generated by such trips.

## 7.3  A Simple Model: Assumptions and Definitions

Consider a trip as a set of linked stops.  The following assumptions are needed: (1) Trips entail at least one stop and a return; i.e. the shortest trip leaves home, stops for a purpose, then returns home. (2) Trips are assumed to return home after a finite number of stops. (3) Trips terminate at home. (4) It is assumed that each stop on a trip is made for one of a given range of purposes only. (5) At every stop there are known or observable probabilities of continuing to other purposes, or of returning home.  These probabilities are not necessarily constant over stops or purposes. This simple view of multistop trips is represented as a decision tree in Figure 7.3.

The concepts and assumptions outlined above can now be used to classify trip types.  A single-stop, single-purpose trip is a trip involving a stop for a single purpose.  A multistop, single-purpose trip chain is a trip involving a sequence of destinations at each of which a stop is made for the same purpose (e.g. home-shop-shop-home).  A multistop, multipurpose trip chain is a trip involving a sequence of destinations at each of which a stop is made for one of a range of purposes (e.g. home-shop-recreation-home).  A much more complex type of representation of trips would involve a sequence of destinations at each of which a stop is made for a number of purposes.  This type of trip is not analyzed here, as the level of complexity in choices is so large as to require a more detailed set of data than is available even in the Hamilton case. The difficulties of collecting micro data about the detailed movements and activities of individuals in the course of a trip to a shopping center, or during a trip to a city shopping district has been cited as a major stumbling block to research on demand driven retail agglomeration (Shepard and Thomas, 1980). Even if the micro data could be gathered, it is not obvious how these observations could be integrated into an aggregate picture of travel behaviour, since so many idiosyncracies of the individual and so many unique circumstances influence any particular trip. In the context of the present analysis, the survey definition of what constitutes a separate stop is important. Trip makers were asked to report new activity as a separate stop if a change in location of more than two blocks was involved.

**Figure 7.3   Conditional Probability Model for Transitions**

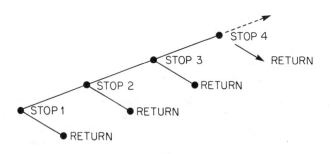

Before proceeding with further definitions, the Hamilton data are used to construct a conditional probability matrix, called $R(m,m-1)$. This matrix contains the conditional probability that a trip which stopped for purpose g on stop m-1, stops for purpose h on stop m. The data for these conditional probabilities use some information from the household travel survey. Tables 7.1-7.4 show numbers (and proportions) of transitions from purpose i on stop (m-1) to purpose j on stop m, (for values of m = 2,3,4,5). In this example there are five possible purposes for a stop: grocery shopping; non-grocery shopping; social and recreational (includes miscellaneous other purposes); work; and return home. These tables are interpreted as follows. Consider Table 7.1 which shows the proportion of first stops for purpose i which stopped for purpose j at the second stop. (Throughout these tables, i indexes rows and j indexes columns.) The first row of Table 7.1 shows the proportion of those making a grocery shopping stop on stop 1 who made (i) a grocery shopping stop on stop 2 (0.078), (ii) a non-grocery shopping stop on stop 2 (0.119), (iii) a social or recreational stop on stop 2 (0.076), (iv) a work stop (0.007), and (v) a return home on stop 2 (0.721). These proportions are taken to represent the probability of change from purpose i on stop 1 to purpose j on stop 2. Similar interpretations are given to the other elements of Table 7.1. The last element of the first row (i.e. 0.721) shows the probability that a trip which stopped for grocery shopping on stop 1, returns home on stop 2, and is of course larger than any other element in the row, reflecting the high probability that a grocery trip is in fact a single-stop trip. The corresponding elements in last position of the second row shows that while non-grocery trips have a high probability of being single-stop trips (0.524) the probability is considerably less than for any other purpose. Tables 7.2-7.4 then show the transition rates for the trips which actually continue on to second, third, fourth and fifth stops. While every one of these tables emphasises the high probability of returning home at any given stop, the combined effect of those trips which <u>do</u> continue on to later activity is very important, as will now be shown.

To determine the probability of interaction between purpose combinations we have to compute the <u>unconditional</u> trip probabilities. That is, to obtain the unconditional probability for an activity at stop m, the conditional probabilities for all previous purposes are removed by multiplying the appropriate conditional probability matrices $R(2,1)...R(m,m-1)$.

### Table 7.1  Results from Hamilton Travel Diary Survey: Stop 1 to Stop 2

Stop 1 to Stop 2, rows show origin purpose, columns show destination purpose, bracketed term is the row proportion.

|  | GROC SHOP | NON-GR SHOP | S/REC | WORK | RETURN HOME | TOTAL ORIGINS |
|---|---|---|---|---|---|---|
| GROCERY SHOPPING | 154 (0.078) | 234 (0.119) | 149 (0.076) | 13 (0.007) | 1421 (0.721) | 1971 |
| NON-GROCERY SHOPPING | 392 (0.096) | 834 (0.204) | 565 (0.139) | 149 (0.037) | 2139 (0.524) | 4079 |
| SOCIAL/REC and OTHER | 383 (0.039) | 874 (0.089) | 1829 (0.186) | 550 (0.056) | 6210 (0.631) | 9846 |
| WORK | 218 (0.031) | 768 (0.111) | 721 (0.104) | 133 (0.019) | 5095 (0.735) | 6935 |
| STOP 2 | 1147 | 2710 | 3264 | 845 | 14865 | 22831 |

### Table 7.2  Results from Hamilton Travel Diary Survey: Stop 2 to Stop 3

Stop 2 to Stop 3, rows show origin purpose, columns show destination purpose, bracketed term is the row proportion.

|  | GROC SHOP | NON-GR SHOP | S/REC | WORK | RETURN HOME | TOTAL ORIGINS |
|---|---|---|---|---|---|---|
| GROCERY SHOPPING | 92 (0.080) | 109 (0.095) | 116 (0.101) | 8 (0.007) | 822 (0.717) | 1147 |
| NON-GROCERY SHOPPING | 200 (0.074) | 475 (0.176) | 420 (0.155) | 119 (0.044) | 1491 (0.551) | 2705 |
| SOCIAL/REC and OTHER | 129 (0.040) | 312 (0.097) | 839 (0.262) | 250 (0.078) | 1671 (0.522) | 3201 |
| WORK | 40 (0.048) | 89 (0.106) | 269 (0.321) | 46 (0.055) | 395 (0.471) | 839 |
| STOP 3 | 461 | 985 | 1644 | 423 | 4379 | 7892 |

## Table 7.3  **Results from Hamilton Travel Diary Survey: Stop 3 to Stop 4**

Stop 3 to Stop 4, rows show origin purpose, columns show destination purpose, bracketed term is the row proportion.

| | GROC SHOP | NON-GR SHOP | S/REC | WORK | RETURN HOME | TOTAL ORIGINS |
|---|---|---|---|---|---|---|
| GROCERY SHOPPING | 40 (0.087) | 49 (0.106) | 54 (0.117) | 3 (0.007) | 315 (0.683) | 461 |
| NON-GROCERY SHOPPING | 82 (0.083) | 150 (0.153) | 166 (0.169) | 32 (0.033) | 553 (0.563) | 983 |
| SOCIAL/REC and OTHER | 72 (0.045) | 141 (0.087) | 347 (0.215) | 63 (0.039) | 991 (0.614) | 1614 |
| WORK | 21 (0.050) | 39 (0.092) | 93 (0.220) | 23 (0.054) | 247 (0.584) | 423 |
| STOP 4 | 215 | 379 | 660 | 121 | 2106 | 3481 |

## Table 7.4  **Results from Hamilton Travel Diary Survey: Stop 4 to Stop 5**

Stop 4 to Stop 5, rows show origin purpose, columns show destination purpose, bracketed term is the row proportion.

| | GROC SHOP | NON-GR SHOP | S/REC | WORK | RETURN HOME | TOTAL ORIGINS |
|---|---|---|---|---|---|---|
| GROCERY SHOPPING | 14 (0.065) | 17 (0.079) | 18 (0.084) | 1 (0.005) | 165 (0.767) | 215 |
| NON-GROCERY SHOPPING | 19 (0.050) | 42 (0.111) | 72 (0.190) | 18 (0.048) | 227 (0.601) | 378 |
| SOCIAL/REC and OTHER | 22 (0.034) | 58 (0.089) | 155 (0.238) | 34 (0.052) | 383 (0.587) | 652 |
| WORK | 5 (0.041) | 10 (0.083) | 40 (0.331) | 8 (0.066) | 58 (0.479) | 121 |
| STOP 5 | 60 | 127 | 285 | 61 | 833 | 1366 |

The operational definition of the unconditional probability of activity on the $m^{th}$ stop of the trip is therefore as follows:

$$S(2,1) = R(2,1)$$
$$S(3,1) = R(2,1).R(3,2)$$
$$S(4,1) = R(2,1).R(3,2).R(4,3)$$
$$S(5,1) = R(2,1).R(3,2).R(4,3).R(5,4). \tag{7.1}$$

The results of these manipulations are shown in Table 7.5.    As an example of their interpretation, consider the first row of each matrix. The first element in each of these rows shows the probability that a trip which stopped for grocery shopping on stop 1: stops for grocery shopping on stop 2 (0.0781); on stop 3 (0.0184); on stop 4 (0.0074); and on stop 5 (0.0020). The second element of the first row of each of the matrices shows the probability that a trip which stopped for grocery shopping on stop 1 stopped for non-grocery shopping, on stop 2 (0.1187); on stop 3 (0.0363); on stop 4 (0.0128); and on stop 5 (0.0042). Notice that while probabilities of transition decline as the chain becomes longer, certain transitions remain more likely than others. For example, the unconditional probability of transition into non-grocery activity remains higher than that for grocery shopping throughout.

The mean number of transitions into state h on or before the $m^{th}$ stop for a trip which started in state g on stop 1 is computed as:

$$N(m,1) = Q(1,1) + Q(2,1) + ... + Q(m,1) \tag{7.2}$$

where $Q(1,1)$ is an identity matrix, and $Q(2,1)$, $Q(3,1)$,... are the upper left hand 4 by 4 submatrices of $S(2,1)$, $S(3,1)$,... (This result is discussed further in O'Kelly, 1981b.) The matrices needed to compute $N(m,1)$ are shown in Table 7.5 for the small numerical example. The results of applying equation (7.2) to these data are shown in Table 7.6.

Table 7.5  **Unconditional Probabilities of Transition into Purpose in Column on Stop m, Given Stop for Purpose in Row on Stop 1**

| STOP 2 | GROC SHOP | NON-GR SHOP | S/REC | WORK | RETURN HOME |
|--------|-----------|-------------|-------|------|-------------|
| GROC  | 0.0781 | 0.1187 | 0.0756 | 0.0066 | 0.7210 |
| NGROC | 0.0961 | 0.2045 | 0.1385 | 0.0365 | 0.5244 |
| SREC  | 0.0389 | 0.0888 | 0.1858 | 0.0559 | 0.6307 |
| WORK  | 0.0314 | 0.1107 | 0.1040 | 0.0192 | 0.7347 |
| HOME  | 0.0000 | 0.0000 | 0.0000 | 0.0000 | 1.0000 |

| STOP 3 | GROC SHOP | NON-GR SHOP | S/REC | WORK | RETURN HOME |
|--------|-----------|-------------|-------|------|-------------|
| GROC  | 0.0184 | 0.0363 | 0.0483 | 0.0120 | 0.8850 |
| NGROC | 0.0301 | 0.0624 | 0.0895 | 0.0225 | 0.7955 |
| SREC  | 0.0198 | 0.0433 | 0.0843 | 0.0217 | 0.8308 |
| WORK  | 0.0158 | 0.0346 | 0.0538 | 0.0143 | 0.8815 |
| HOME  | 0.0000 | 0.0000 | 0.0000 | 0.0000 | 1.0000 |

Table 7.5 (CONTINUED)

| STOP 4 | GROC SHOP | NON-GR SHOP | S/REC | WORK | RETURN HOME |
|---|---|---|---|---|---|
| GROC | 0.0074 | 0.0128 | 0.0213 | 0.0038 | 0.9546 |
| NGROC | 0.0129 | 0.0226 | 0.0383 | 0.0069 | 0.9193 |
| SREC | 0.0102 | 0.0181 | 0.0325 | 0.0060 | 0.9332 |
| WORK | 0.0074 | 0.0130 | 0.0224 | 0.0041 | 0.9532 |
| HOME | 0.0000 | 0.0000 | 0.0000 | 0.0000 | 1.0000 |
| STOP 5 | GROC SHOP | NON-GR SHOP | S/REC | WORK | RETURN HOME |
| GROC | 0.0020 | 0.0042 | 0.0094 | 0.0020 | 0.9824 |
| NGROC | 0.0036 | 0.0075 | 0.0168 | 0.0036 | 0.9686 |
| SREC | 0.0029 | 0.0062 | 0.0140 | 0.0030 | 0.9738 |
| WORK | 0.0021 | 0.0044 | 0.0098 | 0.0021 | 0.9817 |
| HOME | 0.0000 | 0.0000 | 0.0000 | 0.0000 | 1.0000 |

Table 7.6  **Mean Number of Transitions into State h on or before the Fifth Stop, Starting in State g on Stop 1**

|  | GROCERY | NON-GROC | SOC-REC | WORK |
|---|---|---|---|---|
| GROCERY | 1.1059 | 0.1721 | 0.1546 | 0.0245 |
| NON-GROC | 0.1427 | 1.2970 | 0.2830 | 0.0695 |
| SOC-REC | 0.0718 | 0.1564 | 1.3166 | 0.0866 |
| WORK | 0.0567 | 0.1627 | 0.1899 | 1.0396 |

The interpretation of these data has been used by O'Kelly (1983a) to motivate a model of retail shopping center size incorporating multipurpose trips. The interpretation is based on the following observations: the trip maker's first stop activity is shown in the rows of the matrix in Table 7.6. The entries in the columns of the table show the average number of additional stops at these activities on trips with up to 5 stops. For example given that the first activity on the trip was grocery shopping, the diagonal element of the table counts the one grocery shopping stop, together with the average number of extra grocery shopping stops (1.1059), the number of additional non-grocery stops (0.1721) and so on. Given an initial grocery stop the most likely subsequent stop is for non-grocery shopping; similarly interpreting the columns, the most likely transition into a grocery shopping stop is from an initial non-grocery stop. Moving to the second activity, non-grocery shopping is frequently chained with other non-grocery stops, but also with social-recreational activity. It might be inferred from these data in the Hamilton case that the location of grocery facilities is sensitive to the location of higher order non-grocery facilities but that non-grocery facilities seem to owe a lot of their interaction to trips which start out for other non-grocery activity.

The two remaining categories of travel show an interesting contrast: the social and recreational travel purpose is a very common purpose following all the other categories of travel in this table; by contrast work travel rarely follows other activity. These results must be used cautiously as they are specific to the Hamilton data set; furthermore these aggregate indices reflect the behaviour of people of different ages, both sexes, and various family sizes and residential locations. These factors have important bearings on travel behaviour in the urban context, as is shown in Madden (1980), Howe and O'Connor (1982), and the reviews in Hanson (1986).

As an extension of this simple model O'Kelly (1981b) superimposes a spatially disaggregated interaction model on the purpose transition tables shown here. In that earlier research actual data were used to calculate the spatially disaggregated matrices with the result that a system with 4 purposes and 20 destinations involved the analyses of an 80 by 80 system. This rather unwieldy formulation prompts the following compact version of the model: The major assumption which simplifies the analyses is that transitions between locations obey the same spatial interaction matrix $T$ throughout, but that the purpose transition probabilities are derived from actual data and so may show "fatigue" and other trends in purpose transitions throughout the course of the trip. By making the assumption of constant interaction matrix $T$ with elements $T_{ij}$, the following can be shown to be spatially disaggregated <u>conditional</u> transition probabilities:

$$P(2,1) = T * R(2,1)$$
$$P(3,2) = T * R(3,2)$$
$$P(k,k-1) = T * R(k,k-1) \tag{7.3}$$

where * indicates the Kronecker product. To clarify this notation, suppose that the $R(m,m-1)$ matrix is z by z, and that the interaction matrix is n by n. Thus there are z purposes and n locations in the system. A fully disaggregated approach would require a $(nz)^2$ matrix of values. A compact representation of the system can be put together if we consider the following:

$$P(k,k-1) = \begin{vmatrix} T_{11} \cdot R(k,k-1) & T_{12} \cdot R(k,k-1) & \ldots & T_{1n} \cdot R(k,k-1) \\ T_{21} \cdot R(k,k-1) & T_{22} \cdot R(k,k-1) & \ldots & T_{2n} \cdot R(k,k-1) \\ \cdot\cdot & \cdot\cdot & & \cdot\cdot \\ T_{n1} \cdot R(k,k-1) & T_{n2} \cdot R(k,k-1) & \ldots & T_{nn} \cdot R(k,k-1) \end{vmatrix}$$

where each of the $n^2$ blocks of this matrix is computed by taking the scalar multiple of $T_{ij}$ and the $R(k,k-1)$ matrix. The matrix $P(k,k-1)$ contains the spatially disaggregated conditional purpose transition probabilities between stop k-1 and k and the dimensions of the matrix are (nz) by (nz). The unconditional matrices are then found using an operation like that in equation (7.1) above:

$$S(2,1) = \{T * R(2,1)\}$$
$$S(3,1) = \{T * R(2,1)\} \cdot \{T * R(3,2)\}$$
$$\ldots$$
$$S(m,1) = \prod_{k=2}^{m} \{T * R(k,k-1)\} . \tag{7.4}$$

Now because of the properties of the Kronecker product, it can be seen that we can "extract" the constant matrix $T$ and therefore:

$$S(m,1) = T^{m-1} * \{\prod_{k=2}^{m} R(k,k-1)\}. \tag{7.5}$$

where $T^{m-1}$ is the $(m-1)^{st}$ power of $T$ and the sizes of matrices to be calculated are evidently much more compact under this constant spatial interaction matrix assumption.  To reinforce this result, use i and j to denote locations and g and h to denote purposes:   then the unconditional probability of visiting j for purpose h on stop m, having visited i for purpose g on stop 1 is:

$$S_{ij}^{gh} (m,1) = (T_{ij})^{m-1} (R_{gh})^{m}$$    (7.6)

where $(T_{ij})^{m-1}$ is the $(i,j)^{th}$ element of the $(m - 1)^{th}$ power of the spatial interaction matrix, and $(R_{gh})^{m}$ is the $(g,h)^{th}$ element of the product of $\{R(2,1) \cdots R(m,m-1)\}$. If the spatial interaction transition probabilities vary by stop or "stage" of trip then the full notation in O'Kelly (1981b) should be used. The reader is referred to the aforementioned paper for the derivation of several descriptive statistics from the transition model. These statistics include final demand levels for each trip purpose, the variances of demand levels for activities in each location, and finally the mean and variance of the number of stops on a trip.

## 7.4   Spatial Linkages

So far activity linkages have been discussed in a simplified spatial context.  The more interesting problem involves the spatial linkages which emerge as a result of multistop, multipurpose trips.  The importance of this land use-activity linkage question has been recognised by Hensher (1976), Hanson (1980a) and others.  The fact that trips entail complex spatial interactions and sequences of activities implies that there is a fundamental interdependence between locations and that relative location is probably the basic determinant of spatial flows.  This idea implies that some places receive large inflows to the site, which are not on the first stop of a trip.  Evidence of the importance of this effect can also be shown with reference to some data from the Hamilton survey.  Table 7.7 shows the numbers of single-stop, single-purpose grocery shopping trips to each of the 17 zones, and these are compared with the total number of stops in each zone.  For grocery shopping stops zones 6, 9, 10, 11, 15 and 16 attract over three times as many total stops as single-purpose stops.  In other words, simple home- based, single-purpose trips account for only about one third of all the stops at these locations, and two thirds of the stops at these locations are on multistop, multipurpose trips.  These locations are the Hamilton C.B.D., West, Central and East Mountain and the towns of Stoney Creek and Dundas respectively. Multipurpose shopping trips are important even for grocery shopping trips, especially to town centers (Hamilton, Stoney Creek and Dundas).  Zones 9, 10 and 11 (containing suburban malls) have large ratios of total demand to initial demand, compared with other suburban locations (3, 4, 8).  This result might be explained by either the different socio-economic status of the households (9, 10, 11 are the areas with higher household income), or by the fact that the shopping centers in 9, 10, 11 are intervening opportunities between the suburban fringe and downtown.

Table 7.7  **Characteristics of Trips from Hamilton Data Set**

| ZONE | GROCERY SHOPPING | | | | | NON-GROCERY SHOPPING | | | | |
|---|---|---|---|---|---|---|---|---|---|---|
| | A | B | C | D | E | A | B | C | D | E |
| 1 | 13 | 4 | 17 | 38 | (2.9231) | 19 | 20 | 39 | 56 | (2.9474) |
| 2 | 21 | 4 | 25 | 41 | (1.9524) | 34 | 23 | 57 | 100 | (2.9412) |
| 3 | 116 | 46 | 162 | 269 | (2.3190) | 173 | 130 | 303 | 654 | (3.7803) |
| 4 | 177 | 62 | 239 | 396 | (2.2373) | 234 | 177 | 411 | 833 | (3.5598) |
| 5 | 82 | 30 | 112 | 202 | (2.4634) | 126 | 150 | 276 | 572 | (4.5397) |
| 6 | 139 | 60 | 199 | 430 | (3.0935) | 264 | 256 | 520 | 1143 | (4.3295) |
| 7 | 79 | 13 | 92 | 156 | (1.9747) | 77 | 69 | 146 | 296 | (3.8442) |
| 8 | 56 | 16 | 72 | 157 | (2.8036) | 113 | 160 | 273 | 555 | (4.9115) |
| 9 | 115 | 47 | 162 | 350 | (3.0435) | 99 | 117 | 216 | 414 | (4.1818) |
| 10 | 130 | 64 | 194 | 399 | (3.0692) | 318 | 300 | 618 | 1220 | (3.8365) |
| 11 | 127 | 63 | 190 | 381 | (3.0000) | 118 | 96 | 214 | 395 | (3.3475) |
| 12 | 37 | 7 | 44 | 91 | (2.4595) | 59 | 32 | 91 | 178 | (3.0169) |
| 13 | 32 | 16 | 48 | 89 | (2.7813) | 60 | 39 | 99 | 180 | (3.0000) |
| 14 | 0 | 0 | 0 | 1 | ( - ) | 21 | 16 | 37 | 72 | (3.4286) |
| 15 | 76 | 41 | 117 | 231 | (3.0395) | 108 | 84 | 192 | 377 | (3.4907) |
| 16 | 112 | 45 | 157 | 338 | (3.0179) | 96 | 116 | 212 | 498 | (5.1875) |
| 17 | 80 | 19 | 99 | 198 | (2.4750) | 102 | 77 | 179 | 338 | (3.3137) |
| (18, 19) and 20 | 29 | 13 | 42 | 87 | (3.0000) | 118 | 78 | 196 | 399 | (3.3814) |
| TOTAL | 1421 | 550 | 1971 | 3854 | (2.7121) | 2139 | 1940 | 4079 | 8280 | (3.8710) |

COLUMN A Number of first stops in zone i, for purpose named in header, which were followed by a RETURN HOME; (i.e. single-stop single-purpose shopping trips to zone i).

COLUMN B Number of first stops in zone i, for named purpose, which continued to a further activity.

COLUMN C Combined number of first stops for named purpose = A + B.

COLUMN D Total number of stops for named purpose in zone i.

COLUMN E "Scale factor" showing the ratio of total stops to single-stop single-purpose stops for that purpose by zone. (Column D/Column A)

Source: Based on O'Kelly (1981c) Tables 3.6 and 3.7

There are only limited access routes from the upper to the lower city, and these serve to channel traffic past the upper city shopping plazas. Thus the idea of accessibility to other opportunities again can be thought of as playing a major role in spatial interaction.

Other results of the importance of multipurpose travel to the distribution of demand for non-grocery facilities are illustrated in Table 7.7. Generally the scale factors for total demand as compared with demand occurring on single-stop, single-purpose trips are higher than those for grocery shopping. Single-stop, single-purpose trips account for less than one-third of all the stops in the areas, for non-grocery shopping. Particularly high ratios of total demand to 'single-stop' demand occur for zones, 5, 6, 8, 9 and 16, which are respectively the area east of the Hamilton C.B.D., the C.B.D., the west end of Hamilton, the west mountain, and Dundas. The factors involved are similar to those for grocery shopping. The west end of Hamilton has an unexpectedly high scale up factor (4.9) which is probably due to the heterogeneity of the land uses within the area and its proximity to Dundas and Ancaster. The results above show that the only purpose for which the simple single-stop, single-purpose trip assumption is a reasonable approximation is for work trips. Shopping trips, in this study, and in previously reported studies involve multistop purposes in as many as 74% of the cases. Important spatial and functional linkages are implied by this type of behaviour.

7.5   Models of Multistop, Multipurpose Trips

Several authors have attempted to improve the conventional spatial interaction model to incorporate multistop trip making. For example, Webber, O'Kelly and Hall (1979) define trip patterns, which allow stops at up to two facilities. While the number of possible patterns remains manageable for a small system with less than 20 facilities, the technique does not provide a very rich coverage of trip types and rapidly encounters combinatoric problems. Horowitz (1980) has made a similar point in connection with disaggregate models of trips.

O'Kelly (1981b) provides a model which views a trip as a linked chain, and the model allows non-constant transition matrices and several purposes. That research on multistop, multipurpose trip-making focuses explicit attention on the probability of sequences of activity and deduces measures of the interdependence between places. This model has been developed in a series of papers (O'Kelly, 1983a; 1983b) and an overview of the method was presented in the previous section of this chapter. In essence this framework allows several important aspects of trip making to be incorporated without being subject to the criticisms (Hanson, 1979; Damm 1979; Damm 1980) of earlier markovian models (e.g. Horton and Shuldiner, 1967; Horton and Wagner, 1969; Sasaki, 1972; Wheeler, 1972; Kondo, 1974). First, the probability of making a stop for purpose g following a stop for purpose h is dependent on the stage of the trip at which the stops take place. This modification is achieved using Howard's (1971) ideas on time-varying Markov processes. Second, the model takes into account spatial flows at each stage of the trip by disaggregating the transition probabilities by origin and destination. (A similar technique is used in Sasaki, 1972. An alternative method proposed by Gilbert, Peterson and Schofer (1972) and later by Kitamura and Lam (1981) is to view the trip as a sequence of stops, and to model the transitions between stops as a Markov Renewal model.)

A detailed notation for an accounting framework using these ideas has been used by Borgers and Timmermans (1986), who apply the model to central city pedestrian flow. Their paper takes the notation developed in O'Kelly (1981b) and in turn models the transition probabilities as a function of the total amount of retail floorspace in retail sector g at location j (they call it link j) and the distance separating the potential stops on the trip. They add two other components to the model - namely a route choice model (given that destinations are to be included in the trip, what is the route chosen for the trip ?); and, a model of impulse stops, to account for unplanned activity added to the trip as it progresses.

Each of these sub-models deserves further comment and research: their destination choice model "assumes that the probability of choosing a particular shopping street is proportional to a function of the floor space in that street and inversely proportional to the distance separation between stops and the attractiveness of competing streets" (Borgers and Timmermans, 1986, 122). Thus they use a production-constrained interaction model to represent the transition probabilities in the markov chain. This is a useful contribution in the area of micro-movement within shopping districts (as called for by Shepard and Thomas, 1980). The second component of their model is a route choice prediction based on a fixed set of destinations: it is clear that a fixed set of destinations can be linked together on a trip chain in a number of ways, and that any one of these particular trip chains can be traversed along a number of different routes. These two issues are quite different: the first is a matter of the trip makers ability to organise a sequence of destinations and activities in order to avoid unnecessary travel. Borgers and Timmermans (1986) do not attempt to analyze this question, although O'Kelly and Miller (1984) have set up an empirical study of this type. The second issue is that of taking a fixed set of destinations in the order that they are visited by the trip maker, and assessing the utility associated with various routes between these stops. The authors are able to produce a good fit to their data, using some heuristic rules to eliminate unlikely route choices, much in the same way as Landau, Prashker and Alpern (1982) eliminate infeasible destinations from activity spaces. Finally the Borgers and Timmermans' (1986) paper introduces a model which permits an impulse stop. Again there are shortcomings in the method in that the data are derived from retrospective questioning of individuals as they leave the shopping area, so that it is difficult to be sure that the traveller actually made a planned stop, or whether simply passing an opportunity reminded the trip maker of some activity that needed to be undertaken. These comments indicate some of the major difficulties in attempting to put together a model of complex interaction even when the spatial scope of the activities is relatively confined to a single shopping district (see also Guy and Wrigley, 1987).

## 7.6   Implications for Modelling Interaction

This section of the chapter now changes focus and attempts to summarise efforts which have been made to model the types of data which were reviewed in the introduction. Thus, we are concerned primarily with the implications for model design of some observed characteristics of non-work travel. This is not a catalog of facts of about non-work travel, rather it is a discussion of the difficulties inherent in representing non-work travel (see also Damm, 1982). Further, the review places the data in context by emphasising the importance of multiple purpose travel and some of the difficulties of modelling this behaviour. Non-work travel is important for several reasons. It is an important component of overall travel as can be seen from Tables 7.1-7.4 above. It is discretionary - any travel policy is likely to be most effective in modifying behaviour when there exists flexibility in choice behaviour. Non-work trips are complex: they are not inherently suited to simple home based single-stop analysis, and typically non-work trip behaviour includes travel for diverse purposes over a wide variety of distances; (see the early model of Horowitz, 1978). Conventional notions of distance impedance are not as binding as in the case of journeys to work: individuals may either wish to, or have to, travel long distances (compared with journey-to-work) for some activities.

In most non-work trips the number of discrete alternatives from which to choose is large, especially when the interdependence between mode and destination choice is considered. Constraints of car availability, knowledge, income and mobility may vary in terms of their impact on choice. Furthermore it is difficult to determine motives or behavioural decision rules from revealed preferences. The multiplicity of choices facing the individual (see Horowitz, 1980) is a major problem: when destination, mode, purpose, and time of day choices are compounded, the number of alternative travel packages from which to choose becomes immense (Adler and Ben-Akiva, 1979). The actual travel behaviour is likely to depend on the contents of the activity schedule; the day of the week; and, time of day (Kitamura, 1983; 1984a; 1984b; and Pas, 1982).   In the U.S., Canada and the U.K. non-work travel accounts for peak loads at weekends and in the evening, but cultural and religious differences result in completely different shopping and recreational practices in other countries (Israel for example). A specific example may be helpful here: Davies (1976) in an early study of trip generation found that in Coventry (U.K.) there was a strong tendency for women to shop on Friday while overall traffic and trade peaked on Saturday. A tabulation of comparable results from the Hamilton travel survey is shown below. Table 7.8 shows the number and percentage of home-based grocery, non-grocery and social/recreational trips by day of the week. While both types of shopping activity are dominated by Thursday, Friday and Saturday, social and recreational trips are more uniformly distributed across the days of the week. An analysis of the occurrence of multipurpose trips over the days of the week shows that Thursday, Friday and Saturday account for about 55% of all the multipurpose trips in the sample.

**Table 7.8   Distribution of Home-Based Trips by Day of the Week**

|  | Grocery | | Non-grocery | | Social/Recreational and other non-work | |
|---|---|---|---|---|---|---|
| SUNDAY | 90 | ( 4.8) | 159 | ( 4.5) | 1214 | (15.9) |
| MONDAY | 191 | (10.2) | 455 | (12.8) | 1025 | (13.4) |
| TUESDAY | 222 | (11.8) | 495 | (13.9) | 1124 | (14.7) |
| WEDNESDAY | 248 | (13.2) | 439 | (12.4) | 1110 | (14.5) |
| THURSDAY | 342 | (18.2) | 611 | (17.2) | 1081 | (14.1) |
| FRIDAY | 341 | (18.1) | 724 | (20.4) | 963 | (12.6) |
| SATURDAY | 445 | (23.7) | 668 | (18.8) | 1134 | (14.8) |
| TOTAL | 1879 | (100.) | 3551 | (100.) | 7651 | (100.) |

Source: Authors' calculations from Hamilton Travel Diary Survey.

Discussion of the weekly activity pattern in Israel reveals significant differences between activity patterns on different days of the week (Hirsh, Prashker and Ben-Akiva, 1984). The commercial week in Israel is made up of six workdays (Sunday to 1pm Friday), and stores are closed on Friday afternoon and on Saturday. As a result of the restricted hours on Friday and in anticipation of the closures on Saturday, the shopping rates on Friday morning are roughly twice those of most other mornings (see the detailed data in their Table 3).

It would seem to be difficult to integrate every possible type of travel behaviour into one model; at the very least cross-cultural differences will make such an attempt pointless. On the other hand there seem to be some systematic components to travel behaviour (both simple single purpose trips for groceries at the nearest store, and indeed elaborate tours to find the best bargain): the task facing the model builder is to allow both sufficient realism that actual complex behaviour can be represented, and at the same time an awareness of the underlying logic of the spatial choice process elucidated in Chapter 1. With this difficult goal in mind, we now attempt to survey some implications for model design integrating both multipurpose travel with simple travel. While the introduction of multipurpose trips vastly complicates the description and analysis of non-work travel, it also helps our understanding of many of the otherwise (apparently) non-systematic components of travel. If we define the travel pattern to be the set of trip tours made by an individual or household within a fixed time period, then, how does a household choose a travel pattern? It could choose to minimise travel time by joining all the activities on the same trip, but this is clearly not realistic since only linked trips would emerge and in fact many trips are quite simple. What we need therefore is a framework that allows the coexistence of single and multiple purpose trips. The organisation of the trip may be of little concern to the trip-maker, so the problem should not be artificially classified according to the number of stops. The discussion is organised along the lines of three major problems/themes in modelling. These problems are: (1) definitions of the appropriate choice set; (2) utility function specification; and (3) consideration of the role of search and decision making. These issues arise whether single or multiple purpose travel is considered.

## 7.7    Choice Set Definitions - Activity Spaces for Interaction

A "choice set" is often (mistakenly) identified with the alternatives frequently visited (e.g. Recker and Kostyniuk, 1978; and McCarthy, 1980). In our view this is too restrictive in that the reported centers which are frequently visited are essentially the objects to be predicted in the destination choice model, so that if the choice decision is assumed to be restricted to the centers which are actually visited we have effectively prejudged the answer. On the other hand, considering all the possible alternatives from which choice could feasibly occur, is too wide. There are large numbers of stores/opportunities in cities; (e.g. Artle, 1959) however not all are actively considered, because many locations cannot be reached within a given time allocation for travel (see for example Landau, Prashker and Hirsh (1981); Burns (1979); and Carlstein, Parkes and Thrift (1978)). Here we take choice set to mean a list of discrete alternatives, and the extent of this set should be a compromise between the two extremes above; one operational rule has been proposed which allows the residents of a particular part of a city to choose from the bundle of alternatives which are patronised by those living in the same neighbourhood. (Using this definition, Miller and O'Kelly (1983) report that grocery-shopping choice sets ranged from 2 to 19 with an average of 11 for the residents of the neighbourhoods in Hamilton, Ontario.)  A variety of other deterministic rules for choice set delimitation have been based on the notion of feasible scheduling; an excellent review as well as a novel probabilistic approach is suggested in Swait and Ben-Akiva (1987a; 1987b).

The size and shape of the retail consumer's choice set is of critical importance, and although several authors have called for further analysis of these topics, (e.g. Hubbard, 1978, p. 14) little work has actually been performed.  Hubbard (1978, 5) suggests that much more research should focus on "the explicit inclusion of information relating to the geographic variance or dispersion in retail locations" (he cites Horton 1968, and Openshaw, 1973).  Further, he notes that a methodology might involve an analysis of the variances in the time estimates involved in the individuals' behaviour, and also "by examining variance data relating to the spatial dispersion" of the consumer's feasible retail opportunity set (p. 14), (quoted in Shepard and Thomas, 1980, 66-67).  One well-known regularity is that consumers tend to be more familiar with stores towards the central business district from the home - this idea is sometimes referred to as Brennan's law.  A simple explanation for Brennan's Law can be deduced from work by Moore and Brown (1970) who show that if the density of opportunities is distributed in a circular normal fashion around the city center, then an individual residing at $(R,Y)$ with a circular normal probability of making contact with an opportunity will have the a contact pattern which is displaced towards the city center.   This theoretical result is predicated on the assumption of a circular normal distribution of opportunities for contact.  However, in the case of shopping opportunities there exists some evidence in empirical work by Rogers (1965) that retail opportunities actually exhibit a clustered pattern.  Whatever the actual distribution of opportunities, it is clear from some other theoretical work that several valuable insights might be gained from such aggregate descriptive models. For example: Papageorgiou (1969) derived the negative exponential trip distribution function from a consideration of the density of opportunities. Beckmann (1983) found a similar result. Pipkin and Ballou (1979) used similar concepts in a model of the arrival of trips to the city center.

An interesting aspect of the variation in choice set size by types of activity is the idea that some elements of a choice set may be highly inaccessible from the home base but very accessible from elements of other choice sets. (The most clearly recognisable application of this idea is the opportunity for work-based shopping trips in the CBD.) A primary cause of multipurpose trips is the time saving obtained from travel to clusters of facilities -- the incremental travel time to some otherwise distant elements of the choice set may be very small (see Oster, 1978, and Williams, 1988 on the time savings due to multiple stop trips). The need for a more complex approach to choice set definition is suggested by Adler and Ben-Akiva (1979) who model a selection of travel patterns. They show that complete enumeration of all possible travel patterns is infeasible -- e.g. there exists 2171 possible patterns from 20 destinations with 3 modes and up to 2 stops. Similar experience is reported in Webber, O'Kelly and Hall (1979). The problem of enumerating all the possible combinations of travel patterns is fraught with two major difficulties: first, it is unlikely that individuals examine this number of alternatives (see Chapter 4); second, large choice sets may pose some theoretical problems in that the assumption of "independence of irrelevant alternatives" (IIA) in many choice models is invalid in the presence of "attribute correlation", (Sheppard, 1978). In the case of large numbers of very similar shopping patterns choices, there may well exist several similar stores at similar locations, and these stores may have similar unobserved attributes.

The above criticisms of choice models that assume individuals select destinations from full sets of alternatives, has led to the development of several strategies that account for restricted choice sets.

(1) Since there is evidence that familiar locations are accurately perceived (Golledge and Spector, 1978), a suitable operational method might be to question respondents about their perceptions of travel times, and to include in their choice set all those within prespecified accuracy level.

(2) Landau, Prashker and Alpern (1982) analyze activity constrained choice sets in shopping destination choice modelling. They develop a model which uses knowledge of constraints on shopping travel to generate choice sets of retail locations for consumers. Available time is used to evaluate the feasible alternatives and locations which are infeasible are removed from the choice sets. This procedure yields an improvement of about 5% in the average value of the choice probability in a logit model.

(3) Richardson (1982) considers endogenously generated choice sets to occur as a result of a set of sequential steps. In this model the probability of beginning the search, the number of alternatives searched, and the utility of the selected alternative are related to the cost of each step in the search and the level of prior information possessed by the decision makers (e.g. Burnett and Hanson, 1979). The model must determine the probabilities of alternatives lying in the choice set.

(4) A fourth method for controlling the combinatoric problem of choice set size considers the choice of trip purpose and destination as part of a joint decision in a nested or sequential model (see Kitamura and Kermanshah, 1985). Since we often know from travel surveys where people have gone, this information ought to be used to limit the entire set of possible alternatives to those which are actively considered by the trip makers. A possible mixed strategy might use both random selection of the choice set together with

information on the previously chosen alternatives to construct a composite set. Notice that the choice process is misrepresented if the random choice set is not like the known smaller "actively considered" set. Including a non-valid alternative biases the problem, (further discussion of the choice theoretic foundations of the interaction model is in Chapter 4).

(5) Perhaps the most intricate solution, conceptually, to the problem of selection from restricted choice sets is that of Fotheringham (1986; 1988b) which is discussed in Chapter 4. Fotheringham's competing destinations model is derived from an assumption that individuals make choices from restricted sets of alternatives but that these sets are unknown to the modeller (and perhaps even to the individual). It is possible, however, to incorporate into a spatial choice/spatial interaction framework the probability of an alternative being in the restricted choice set examined by an individual. This probability is shown to be a function of the similarity of one alternative to another (see also Borgers and Timmermans, 1987) which in spatial choice can be measured by a relative location variable.

A difficult area of model design has been highlighted in the discussion of Miller and O'Kelly's (1983) attempt to introduce previous shopping center choices into a destination choice model for the current period (see Pickles and Davies, 1985 and Miller and O'Kelly, 1985). The essence of the Miller and O'Kelly (1983) model is that the utility of a particular destination choice can be represented a function of observed attributes of the destinations; dummy variables indicating previous choices; and unobserved components of utility distributed independently over individuals and time periods. By ignoring a "time constant omitted variable" Miller and O'Kelly (1983) oversimplified the problem, since it is highly likely that choice of shopping trip destination can depend on specific shopper-destination interactions. For example individual shoppers may have particular preference for a center as a result of highly personal past experience, or on the basis of quality and price differences between centers. To the extent that these interactions can be quantified, they should be incorporated into the systematic portion of the utility function. However, because of the difficulty of actually specifying these effects we have to rely on a random disturbance term which varies across shopper-destination combinations but which is constant over time. One of the major contributions of the recent literature in longitudinal data analysis has been the derivation of methods for estimating these more elaborate models (see Davies and Crouchley, 1985).

## 7.8  Utility Functions for Non-Work Travel

Utility functions for retail destination choice were described in some detail in Chapter 6. While that analysis provided a strong example of the spatial interaction approach to supermarket choice, it needs to be reconsidered in the light of the evidence of multipurpose travel behaviour.

An important development in utility theory has been the extension of the exogenous variables to include the interrelationship between a particular trip segment and other trip end activities. It has been clearly recognised that trips often involve multiple activity types and that trips do not take place in isolation from others. Simplistic attempts to account for travel with multiple purposes include Recker and Kostyniuk (1978) who divide shopping locations into those which are more likely to produce multiple activities (e.g. supermarkets in shopping plazas) and those which are more likely to be visited in isolation (e.g. free-standing supermarkets, neighbourhood markets). It is clear that they are simply considering multiple purchases at the same location and not the more realistic trip chain (tour, multi-destination, multi-stop, multi-purpose...). They found that their model performed less accurately in trips to the former type of center. The reality of spatial interaction is that activity is often combined on trips; it is clear that combining activity is efficient in a deterministic economic sense (e.g. Bacon 1984; Mulligan 1983; Papageorgiou and Brummell 1975; Adler and Ben-Akiva, 1979; Damm and Lerman, 1981; Ghosh and

McLafferty, 1984; and Kitamura, 1984b).  More interesting perhaps is the set of behavioural ideas behind combinations of trip purposes.  Baumol and Ide (1956) model multistop trips as part of a search for goods.  Shoppers make additional stops on a trip until some satisfaction level of utility attained.   This idea is taken up in more recent research search/decision literature (see for example Richardson, 1982; and Hay and Johnston, 1979).

It is clear that different activities are combined with each other in various different ways at different times because of: (a) the differential rates at which demands accumulate; (b) scheduling considerations; (c) spatial constraints; and (d) the variations in accessibility to opportunities from particular locations within the choice set.  The problem can be stated as follows: households accumulate demand for non-work travel as a result of depletion of stocks, the need to service passengers (e.g. school children), the need for recreation and social visits and miscellaneous other activities in the daily routine.  Because of the joint nature of many of these travel demands there exists a strong interdependence between various components of the daily activity schedule.  The household trades off the desire to meet needs, with the cost of fulfilling them.  Although there really is not a strong "theory" of what motivates/drives household decision making,  Adler and Ben-Akiva (1979) examine the predictability of household travel pattern choice based on the characteristics of the travel patterns.  They argue as follows:  travel patterns with large numbers of individual stops would be "good" because highly separated activities are more likely to be more easily scheduled and are more likely to take place at the most desirable location and time. The utility of travel is represented as a function of four major variables. (1) S: scheduling convenience (utility is positively related to the degree to which patterns fit into the routine). (2) T: level of service (utility is inversely related to travel costs "expenditures"); (3) D: attributes of destination (utility increases with the "appropriateness" of the destination). (4) S: socio-economic characteristics of the household.

Their expectations about coefficient signs for some of the variables are relatively weak. The idea that utility is gained from doing things separately gives more weight to the ease of scheduling than to the gains from combining a series of possibly undesirable activities on the one trip -- that is they ignore the possibility that there may be utility gains from accomplishing several tasks in the one trip.  That there is such a motive behind some travel is borne out by the ability of malls to attract "one-stop" shoppers.

Generalisations about socio-economic impacts on household trip making have been one of the major contributions of travel demand analysts. While a vast array of data has been summarized in Damm (1983) and Daniels and Warnes (1980), the purpose here is simply to outline some selected results, and to tie these results back to the Hamilton Survey. The results are primarily discussed in terms of shopping activity. The literature contains a number of relevant reports of variations in shopping trip frequency across socio-economic categories. One of the most commonly reported effects is that relating to household size: Robinson and Vickerman (1976) found that the number of shopping trips in a fixed period is positively related to household size. MacKay (1973) also hypothesised that size of household is important but in fact he found that the number of children in the family is the most significant size related variable. The impact of household income on trip propensity is also widely studied, although some conflicting empirical results have been found. Robinson and Vickerman (1976) find a positive relationship while Daws and Bruce (1971) [cited in Daniels and Warnes, 1980] could not establish a simple relationship between the two variables. On the other hand, Davies (1976) found more trips to be generated by lower socio-economic groups and MacKay (1973) shows that lower income households in his sample made a semi-weekly shopping trip in contrast to the lower frequency for high income groups. (See further discussion below.) Many studies that include an indication of female participation in the labor force reveal the great significance of this variable (Johnson and Hensher, 1979; MacKay, 1973). Evidently, working women make fewer separate shopping trips because of time constraints. This result helps to clear up some the ambiguity in the income effect: a two worker household might be expected to have higher income but fewer shopping trips than a one worker household on average, because of the scheduling constraints imposed by the two sets of work journeys. (A specific example is given in Preston and Takahashi (1983).) Therefore, if a study fails to control for the numbers of workers in a household, there will be some conflicting forces loading on the income variable. In addition to these simple results, limited evidence on a more subtle question can be gleaned from the Hamilton Survey: while all income groups reveal a propensity to make multistop, multipurpose trips at the weekend, this tendency is most marked for higher income households. One possible mechanism here is that some higher income households have two workers and week day scheduling constraints force the accumulation of demand for shopping and other activity until the weekend.

A more recent attempt to integrate a households intra-urban trip making with its relative location and socio-economic composition is given in Williams (1988), using the Hamilton Survey as a data base.

## 7.9   Consumer Behaviour: Search Among a Set of Alternatives

Hay and Johnston (1979) model consumer search among a set of alternatives as a function of the benefits of search (potentially finding a location with lower prices) and the costs of search which are proportional to the length of the search and reflect such factors as travel and effort. One of their models deals with complete search amongst a set of alternatives where the alternatives have varying utility. They present ancillary evidence to the effect that prices (and hence utilities) varied among eight supermarkets in central Sheffield in July 1978, (over a short time span). It appears that shoppers make irrational decisions when they purchase goods from a variety of different stores. The postulates on which Hay and Johnston (1979) base their analysis are simple: uncertainty exists (e.g. uncertainty on price variations); consumers reduce uncertainty by searching; search involves costs; search may be curtailed so that the decision may not be optimal. Notice that the costs and benefits of search must vary with the price/type of good. Expensive, infrequently demanded items are worth searching thoroughly for. The costs of such search may be small - in the U.S. the yellow pages together with flat rates for local telephone calls reduce the necessity of travel. Hay and Johnston (1979) summarise results which empirically support the hypothesis that costly searches will be longer than optimal while inexpensive searches will be shorter than optimal. One of the central contributions of their paper is the introduction of a simple model which generates dispersion in destination choice. The idea is that alternatives in the choice set may have attributes which are randomly distributed. The example worked out posits a uniformly distributed random variable for the attraction of each of three destinations where there are known minimum and maximum attraction levels. Suppose A's attraction is distributed on an interval   such that alternative A would strictly dominate alternative B and C. Then of course there would seem to be no rationale for dispersion. However, suppose that this is not the case. Hay and Johnston (1979) show the effects of changes in the variance of the attraction terms. For a given spread of mean attraction levels among three alternatives, a reduction in the variance of attractions enhances the probability of choosing the highest ranked alternative. For a given amount of variance in each attraction level, smaller differences between the groups obviously reduces the probability of choosing the highest ranked alternative. While Hay and Johnston (1979, 798) avoid the issue of actual estimates of attraction distributions, it is possible to speculate on some useful tests and further work: it would be interesting to know how the mean and variance of perceived attraction levels vary with distance of a center from the consumer. Familiarity with local alternatives would presumably give them high attraction levels with low variances. More distant centers might have lower perceived attraction but larger variance. A somewhat circular test of this hypothesis might be constructed using a logit model estimated from revealed preferences. The idea of some centers dominating others could be used to define a choice set. Certain very distant facilities might have such low perceived attraction as to be eliminated entirely from consideration. (e.g. Landau, Prashker and Hirsh, 1981). Alternatively a questionnaire survey could elicit evidence on both attraction variation and choice set composition by asking consumers to either state their level of attraction to a center or to indicate their lack of familiarity with it.

## 7.10  Conclusions

This chapter has reviewed complex non-work interaction from a number of complementary viewpoints: (i) the extent of multipurpose trips as a component of intra-urban interaction set the stage for a commentary on the difficult task of modelling such interactions. (ii) A simplified version of an earlier model (O'Kelly, 1981b) was presented and that model can be seen in applications by O'Kelly (1983b) and Borgers and Timmermans (1986). The review has been couched in terms of modelling enhancements, but there is an underlying concern for broader issues in the transportation planning process. In other words, the main focus of the chapter remains on the estimation of travel demand and the derived patterns of route, destination and time-of-day usage of transportation facilities. The implications for the other component of travel demand (mode choice) has been left for future research.

# CHAPTER 8

---

## APPLICATIONS TO LOCATION-ALLOCATION MODELS

### 8.1 Introduction

Interaction models and location-allocation methods represent two powerful techniques available to the spatial analyst. This chapter reviews the need for a connection between the two methods from both theoretical and applied perspectives; formulates a consistent optimisation framework to represent an interaction based location-allocation model; and, outlines a solution procedure. As an example, the model is applied to the problem of siting 20 facilities among 181 neighbourhoods under a variety of impedance parameters in a simple interaction model. The results show that the proposed method provides interesting location and allocation patterns which exhibit intuitive features: when the impeding effect of distance on interaction is small, facilities are clustered in the center of the region; however, as the model is driven towards the deterministic case by increasing the impedance parameter the model shows more decentralised facility locations. This chapter suggests a formulation for a probabilistic location-allocation model which allows for different distance impedance terms across hierarchical levels of service. In this model, large(small) distance impedance terms imply decentralised(clustered) locational configurations. Appropriate choices of distance impedance terms for two service levels can produce nested location patterns without imposing additional siting constraints.

### 8.2 Delivery Systems and User Attracting Systems

Leonardi (1980) distinguishes between "delivery systems" and "user attracting systems" in location-allocation problems. Since delivery systems are operated entirely by one decision maker, it is safe to assume that all aspects of the problem obey cost minimisation criteria. In general this implies that each demand point is serviced by its nearest facility. In user attracting systems, location decisions are taken by a public authority, but users are free to choose among the available locations and allocate their demand to these facilities on the basis of cost, travel time, or service characteristics. It is known that there will be some stochastic component to this allocation mechanism, and several authors have begun to analyze this case (Batty, 1978; Mirchandani and Oudjit, 1982). In the case of those services which are travelled for by consumers, their decision making can produce a variety of allocations which differ from the simple nearest center rule. Thus, while conventional location-allocation models such as the p-median model (see Hansen, Peeters and Thisse, 1983) possess the property that consumers travel to the nearest center to obtain a service or commodity, individuals often travel further than necessary to purchase goods, because, for example, other goods can be purchased at the same time (see Ambrose, 1968; Toyne, 1971; and Fingleton 1975, and the previous chapter for further discussion of multipurpose travel). The reaction of the location modeler to these observations is perhaps typified by the effort of Weaver and Church (1985) - they develop an interesting variant on the p-median problem which allows a vector of assignment probabilities. These probabilities define the fixed fraction of the demand at node i which is assigned to the nearest, second-nearest, third-nearest etc. facility. Because the assignment proportions do not vary with the facility locations, a modification of conventional optimisation tools is adapted to solve this problem. The authors present examples of up to 55 node problems which are solved efficiently. While their model is an interesting extension of conventional location-allocation models they do not really tackle the stochastic allocation of demand to facilities, based on the relative location of those facilities.

The implications of stochastic allocation of demand to facilities have had a long-standing appeal for theorists, (see for example Tapiero, 1974; Webber, 1978; Mirchandani and Oudjit, 1982; Beckmann, 1983). Probabilistic allocation of demand to facilities implies that expected usage levels can be found but these levels may be more influenced by the attraction of the facility than by its relative location. The extent to which users from outside a proximal area utilise a facility is dependent on the rate at which the probability of interaction declines with distance. If the probability of interaction is greatly diminished by a unit increase in distance, then, travel across jurisdictional boundaries will be slight. If, on the other hand, interaction is relatively unimpeded by distance there will be substantial overlap between jurisdictions and, in fact, some accessible facilities may attract users from many areas. Such a situation would show a large excess of visits to the facility over and above the number of trips originating near the facility, with a resultant problem of allocating costs for the service. This issue has been addressed by Linthorst and van Praag (1981) who model the interdependence between interaction patterns and service areas in the Netherlands.

More theoretically oriented questions can also be pursued in connection with the spatial interaction based location model. The nearest center allocation rule as a special case of the spatial interaction model has been used by Evans (1973), Wilson and Senior (1974) and Webber (1979, 250-280) to show the connections between probabilistic allocation and linear programming. The essence of this result is that as the impedance parameter in an interaction model is increased in absolute value, the negative effect of distance on interaction becomes more pronounced, and as the parameter tends to infinity, the interactions revert to a nearest center allocation. In practice only moderately large values of the parameter are required to achieve this result. For example in the Hodgson (1978) doubly constrained interaction location-allocation model a $\beta$ parameter of 1.5 was generally large enough to give interaction model results similar to the p-median model. Obviously the magnitude of the $\beta$ parameter needed to drive the interaction model towards the deterministic case must depend on the scale of the distances in the system, and on the minimum feasible travel cost needed to satisfy the interaction constraints.

A second theoretically motivated variant on the problem comes from formulating the model in continuous space: Beaumont (1980) extending the work of Williams and Senior (1977) shows that the continuous probabilistic location-allocation problem reduces to the famous Hotelling (1929) ice-cream vendor location model. Webber (1979, 327-340) also used a similar model to obtain the Hotelling agglomeration result under less restrictive assumptions.

These comments on the importance of the role of the allocation mechanism in the location of both private and public facilities form the theoretical motivation for the rest of the chapter. In addition to the points raised so far, there are also several interesting empirical issues raised by the spatial interaction based location-allocation model, as will now be shown.

## 8.3   Discussion of Some Formulations and Applications

The problem of finding solutions to location-allocation models which have imbedded spatial interaction equations is known to be very difficult. Recent contributions in this area have been made by Erlenkotter and Leonardi (1985), O'Kelly (1987a) and Jacobsen (1987). The Erlenkotter and Leonardi (1985) paper proposes a non-linear programming approach. They report sample problems up to 69 nodes. The Jacobsen (1987) effort proposes a DROP heuristic which attempts to eliminate facilities while maintaining a fixed amount of dispersion in the destination choices. Only very small problems are tackled. O'Kelly's procedure has been applied to a 181 node problem and this model will be described in more detail below.

In the majority of cases, spatial interaction based location-allocation models have been solved either by explicit enumeration or through some variant of the Teitz and Bart (1968) interchange heuristic.  Small scale problems are described in Hodgart (1978).  Beaumont (1981) uses a variant of the Cooper (1963)   alternating heuristic to suggest a solution procedure for the planar location-allocation model with a built-in spatial interaction model. Leonardi (1983) has reported some numerical experience with a set of powerful heuristics that exploit the sub-modularity of the objective function.  Probabilistic spatial interaction models have also been solved with heuristic methods by Hodgson (1978), who compared the results of a doubly constrained interaction model with endogenous facility location to a conventional p-median model.  Further work by Hodgson (1984; 1986; and 1987) has modelled facility location and patron behaviour on the basis of a facility's hierarchical level. These efforts have motivated by the observation that the conventional p-median model is unrealistic in its assumption of nearest center allocation and in the lack of attention to the benefits from visiting a high order center even for lower order services. (These issues will be discussed more fully in the application section of this chapter.) A recent novel contribution in this area has been Allard and Hodgson's (1987) demonstration of applied GIS (geographical information system) techniques to the problem of portraying the results of a probabilistic allocation of demand to centers.  The need for such a contribution should be apparent as soon as the analyst realises that a map of allocations must show many flow lines of varying widths to represent the proportion of demand flowing to the various centers from each origin zone.

A large scale programming effort has been contributed by Goodchild (1984) in developing the ILACS geographical information system for facility location.   This model can incorporate the concepts of spatial interaction theory; realistic barriers to interaction (such as rivers); and competition between different groups of facilities (e.g. as a result of franchise competition).

Some limited numerical evidence as to the performance of these models exists.  Goodchild and Booth (1976) devised a probabilistic interaction model to describe the flow of population to 11 swimming pools in London, Ontario. Using this allocation model two new locations were chosen to minimise the total distance travelled by the users of the system. The results of the analysis indicated some discrepancy between the municipality's plans for a new site and their estimate of the optimal location.  Two important features of their analysis were: (i) a realistic estimate of demand based on age profiles throughout the city; and (ii) a demonstration that despite the inclusion of an elastic demand model, the usage of the system seemed to be independent of the supply of facilities.  A more complex example of the interaction between facility location and the demand generated by residential locations is provided in Webber, O'Kelly and Hall (1979).  In their model both residential and facility location are endogenous.  Further, the behaviour of consumers is reflected in the model through a large number of allowable trip bundles; these bundles could contain 0, 1 or 2 visits to facilities.  In the application to data from Hamilton, Ontario, the model of consumer allocation proved to be relatively insensitive to facility location patterns.  Both the Webber, O'Kelly and Hall (1979) and the Goodchild and Booth (1976) applications suggest that an inelastic demand assumption is a good first approximation; that is, it seems to be reasonable to assume that the actual level of demand for the service does not vary with the location of the facilities and that only the observed dispersion of travel among facilities needs to be modelled. The validity of this observation hinges on two assumptions: the given constraint structure of the models and the particular data values used in the calibration.  More generally this model seems to be most appropriate in the case when the service has a fixed demand which must be filled regardless of the location of facilities. For example, Hodgson (1981) was concerned with automobile license renewals which are fixed in total level of demand regardless of the position of the central facility. Of course there are many empirical studies showing apparently lower levels of usage of a service when the service is hard to reach (see the comprehensive treatment in Joseph and Phillips, 1984). In cases where the inelastic demand assumption is invalid (and therefore there is a detrimental effect of inaccessibility on usage <u>and</u> a distance decay effect), other strategies must be employed. This chapter does not address this more general problem although the conversion of the model to an elastic version seems to be relatively straightforward (Beaumont, 1981; Sheppard, 1980).

## 8.4  Spatial Interaction Based Location-Allocation Models

This part of the chapter presents Wilson's production-constrained spatial interaction model as an approach to stochastic allocation of demand to facilities. The model is modified to incorporate integer variables denoting facility location. A specific probabilistic allocation rule for the amount of flow between pairs of locations i and j is written as a function of service demand which originates at i, the attraction at j, and the intervening distance. This classical gravity formulation allocates flows between pairs of locations depending on their characteristics and their spatial arrangement. The idea has been used in such widely known models as Reilly (1931), Huff (1963) and Lakshmanan and Hansen (1965). Recall from Chapter 2's derivation of the production-constrained model that the allocation of a fixed level of demand in each of m zones to a set of facilities in n destinations is formulated as:

$$S_{ij} = A_i O_i w_j \exp(-\beta c_{ij}) \ , \ i=1,...,m; \ j=1,...,n \tag{8.1}$$

where the terms $A_i$ are:

$$A_i = 1/\sum_j w_j \exp(-\beta c_{ij}) \ , \tag{8.2}$$

where $S_{ij}$ is the proportion of demand at i serviced by the facility at location j, $O_i$ is the proportion of demand originating at i, $w_j$ reflects the attraction at j, and $c_{ij}$ is a measure of the spatial separation between i and j. The parameter $\beta$ is either calibrated to match some known interaction data or is defined exogenously. Note that $A_i$ ensures that the sum of all the outflows from origin i add up to the amount of demand at that location. This simple model generalises the all or nothing assignment of demand typically embodied in location models. In this model it is assumed that the demand at zone i is allocated among a set of "open" facilities with known attraction levels $w_j$. Assuming that all the $w_j$ values are positive, the model effectively assumes that facilities exist at all the destinations indexed by j and assigns some fraction of the demand at i to each open destination. In the rest of this section the probabilistic allocation model is modified by representing the presence or absence of a facility at a specific location by an integer variable $Y_j$ which has the value of 1 if a facility is present at zone j and is zero otherwise. For the sake of simplicity no other weight will be associated with each destination, although it easy to generalise the model to represent both the presence/absence of a facility and its attraction. The discussion focuses on the formulation, properties, solution, and application of models of location with a spatial interaction base.

A production-constrained spatial interaction model with endogenous facility location can be derived as the solution to the following optimisation problem:

$$(P1) \quad \underset{(S,Y)}{MIN} \quad (1/\beta)\sum_j Y_j \sum_i S_{ij} \ln (S_{ij} - 1) \ + \ \sum_i \sum_j Y_j S_{ij} c_{ij} \tag{8.3}$$

$$SUBJECT \ TO \quad \sum_j Y_j S_{ij} = O_i \quad i=1,...m \tag{8.4}$$

$$\sum_j Y_j = p \tag{8.5}$$

$$\sum_i S_{ij} \leq Y_j \qquad j=1,...n \qquad\qquad (8.6)$$

$$Y_j \in \{1,0\} \qquad j=1,...n \qquad\qquad (8.7)$$

where the choice variables include $Y_j$, the facility locations. As a guide to interpretation of this model, the following paragraph discusses the objective function (8.3), and the constraints (8.4)-(8.7) in the order that they appear in the model. The objective function in (8.3) is a variant on the simple information minimising model of Chapter 2 (see equation (2.14) and section 2.3.2). The "-1" inside the pair of parentheses adds an irrelevant constant to the objective value and simplifies model notation later. The major differences between (8.3) and equation (2.14) are as follows: the objective is expressed as a function of $S_{ij}$ which is assumed to be scaled to sum to 1 over the values of i and the values of j for which there is an open facility. The overall objective in (8.3) is the negative of a quantity called consumers' surplus (Wilson, Coehlo, Macgill and Williams, 1981, 171) and therefore may be thought of as a measure of consumers' disbenefits from having to travel to spatially dispersed destinations. Before interpreting the first part of (8.3), note its second component is obviously a measure of average interaction costs in the location-allocation system. Indeed, if the allocations are to nearest facilities, then this second component is identical to that used to measure transportation costs in the p-median deterministic model. Note also that this interpretation is entirely consistent with the observation that as the $\beta$ parameter tends to infinity, the allocations are to nearest centers and the first part of the objective will obviously drop out. Therefore, the model collapses to exactly the p-median objective as the impeding effect of distance increases. The first part of (8.3) on the other hand can be seen as the negative of a measure of locational benefits; that is it is a measure of disutility associated with dispersion of choices among the destinations (see Wilson, Coehlo, Macgill and Williams, 1981, 171). The first and second components of (8.3) combine to form a measure of consumer disbenefit which, when minimised by choice of $Y_j$ and $S_{ij}$, produces a set of optimal facility locations. Another way of viewing (8.3) is to see it as a relaxation of the average trip length constraint in the program of section 2.3.2. "Relaxation" in this context means that the parameter $\beta$ is given a fixed value and therefore the value of $\beta.C$ in the lagrangean is a constant and so can be removed from consideration. (C in this context is the sum of system wide interaction costs.)

The constraints (8.4)-(8.7) are simple: in (8.4) the sum of the outflow from each origin to all of the open facilities must equal some known level of demand, namely $O_i$. In (8.5) the sum of the zero-one variables $Y_j$ is constrained to be p, thereby ensuring that exactly p facilities are opened. Constraint (8.6) prevents any allocation from being made to a destination which does not have an open facility. That is the sum of inflows to destination j can be positive only when $Y_j = 1$. Finally (8.7) enforces the integrality condition on the location variables.

The resulting model is of the form:

$$S_{ij} = A_i O_i Y_j exp(-\beta \ c_{ij}) \qquad\qquad (8.8)$$

$$A_i = 1/\sum_j Y_j exp(-\beta \ c_{ij}) \quad i=1,...m \qquad\qquad (8.9)$$

## 8.5  Algorithmic Development

Define **J** to be the set of locations with $Y_j = 1$ (that is, the facility sites). For given **J** the model has the property that the interactions $S_{ij}$ satisfy the origin constraints. Further, suppose that a known impedance factor $\beta$ characterises impeding effect of distance on interaction. Let the travel time from origin zone i to the facility at j be $c_{ij}$. Flows from origin zones to facilities are modeled as shown in equations (8.8) and (8.9), or combining:

$$S_{ij} = O_i Y_j \exp(-\beta \, c_{ij})/(\Sigma_j \, Y_j \exp(-\beta \, c_{ij})) \tag{8.10}$$

Consider again the problem (P1) outlined above; i.e. minimise (8.3) subject to (8.4), where (8.5), (8.6) and (8.7) will be handled implicitly below. Form the lagrangean

$$L = (1/\beta)\Sigma_i\Sigma_j Y_j S_{ij}(\ln S_{ij} - 1) + \Sigma_i\Sigma_j Y_j S_{ij} c_{ij} +$$
$$+ \; \Sigma_i \, a_i \, ( \; \Sigma_j Y_j S_{ij} - O_i), \tag{8.11}$$

then relax the constraint (8.4) by assigning arbitrary values to the multipliers $a_i$ to give the following lagrangean:

$$L' = (1/\beta)\Sigma_i\Sigma_j Y_j S_{ij}(\ln S_{ij} - 1 + \beta c_{ij} + \beta a_i) - \Sigma_i a_i O_i \; . \tag{8.12}$$

Then for all $Y_j = 1$

$$\partial L'/\partial S_{ij} = (1/\beta)(1 + \ln S_{ij} - 1 + \beta c_{ij} + \beta a_i \, ) = 0 \tag{8.13}$$

$$S_{ij} = \exp(-\beta \, (c_{ij} + a_i)) \; , \tag{8.14}$$

and substituting back into $L'$

$$L' = (1/\beta) \; \Sigma_i\Sigma_j Y_j S_{ij}(-1) - \Sigma_i \, a_i O_i \tag{8.15}$$

which is the dual of the original problem. The minimisation with respect to $Y$ can be accomplished by selecting the p smallest values of

$$P_j = \Sigma_i(-1) \exp(-\beta \, (a_i + c_{ij})) \tag{8.16}$$

and then by setting the corresponding values of $Y$ equal to one. This implicit treatment of the constraint on the number of facilities is suggested by Narula, Ogbu and Samuelsson (1977) and is also used in Handler and Mirchandani (1979). The dual problem can thus be written :

$$(D1) \quad MAX \quad FD = (1/\beta) \; \Sigma_j \, Y_j P_j \; - \; \Sigma_i \, a_i O_i \tag{8.17}$$

where the choice variable is the vector of multipliers $A$. The problem is to find the vector $A$ which solves (D1). As a preliminary result, note the following characteristics of the vector $A$. For a given set of facility locations the optimal multipliers can be obtained analytically. Note that

$$\Sigma_j Y_j S_{ij} = O_i \; , \tag{8.18}$$

and so

$$\Sigma_j Y_j \exp(-\beta \, (a_i + c_{ij})) = O_i \; , \tag{8.19}$$

therefore

$$a_i = -(1/\beta) \ln (O_i/( \; \Sigma_j Y_j \exp(-\beta \, c_{ij})) \; . \tag{8.20}$$

It can easily be shown that for a given value of $\beta$, a lower (upper) bound on $a_i$ can be found by setting facilities open in the p largest (smallest) entries in row i of $c_{ij}$. Next observe that for arbitrary values of $A$, and their associated set of facility locations, the primal constraints on the sum of the outflows from i may not be satisfied. Therefore following Narula, Ogbu and Samuelsson (1977), Handler and Mirchandani (1979), Held, Wolfe and Crowder (1974), and Kennington and Helgason (1980), a subgradient search is employed for **A**. The primal constraint (8.18) requires that

$$f_i = O_i - \sum_j Y_j S_{ij} = 0 \tag{8.21}$$

and an appropriate subgradient search sets the $(m+1)^{st}$ value
of $a_i$ as

$$a_i(m+1) = a_i(m) - q_i (f_i) \tag{8.22}$$

where $q_i$ is a step-length parameter determined by experiment, (O'Kelly, 1987a). Notice that if $f_i > 0$, the current flows fall short of the required origin constraint, so that $a_i(m+1)$ is adjusted downward according to (8.14) and from the definition of $S_{ij}$ in (8.22), the flows originating at i increase. Conversely, if $f_i < 0$ , $a_i(m+1) > a_i(m)$ and the flows originating at i are decreased. Of course, if $f_i = 0$ for any i the primal constraint is satisfied and no adjustment is made to the corresponding multiplier. The procedure, in essence, is a method of determining the optimal facility locations and flows by adjusting the set of undetermined multipliers for the demand constraints in such a manner that these "dual" variables maximise the objective D1. In the next section, two examples of location-allocation models based on an interaction equation are be described, and some sample numerical results are provided.

## 8.6 Results

This section reports some numerical results from an application of the algorithm. A 181x181 travel time array from the Hamilton metropolitan area is used as the basis for a test of the algorithms performance. A set of 20 facilities is located in the 181 zones under a variety of characterisations of travel behaviour. More details concerning the Hamilton study area can be found in the chapter on multipurpose travel.
The resulting locations and allocations for $\beta=0.025$ are shown in Table 8.1 while similar data are presented in Table 8.2 for $\beta=0.2$. These location-allocation tables are found using the second step length method in O'Kelly (1987a). The first column shows the location of the twenty facilities, and these locations are shown in Figures 8.1 and 8.2. (These codes are in the format 'zznn' where 'zz' is the zone identifier and 'nn' represents the neighbourhood within that zone.) The inflow to each facility is the sum of the demands allocated to the center. The '% NEAR' column shows the percentage of the demand allocated to the facility, for which this is the nearest alternative. (The maps use heavier shading for the facilities which are nearest centers for a large portion of their users.) The column labelled TIME reports the average travel time (in minutes) for the users of each facility. Spatially, the two systems are quite different, with the small impedance parameter associated with a clustered set of facility locations (Figure 8.1), and the large parameter showing a more dispersed set of locations (Figure 8.2). The major contrast between the two tables is that the system with $\beta=0.025$ exhibits a very uniform set of facilities with little or no preference for the nearest facility. On the other hand, the system with a larger impedance parameter ($\beta=0.2$) displays greater dispersion in facility locations as well as more marked use of the nearest facility and therefore shorter average access times.

## 8.7 Application: Service Delivery in an EMS System

The location of depots and the allocation of equipment in emergency medical service (EMS) systems usually are planned by the same central decision maker. Such services involve the selection of a site for emergency vehicles and a dispatcher who allocates the appropriate vehicle to the emergency site. In the usual case, the nearest center will have a vehicle available and the dispatcher will allocate this vehicle to the emergency. However, there are three sets of complicating factors which will frequently result in a vehicle other than the nearest being dispatched: (1) the nearest center may not have a readily available unit; (2) the nature of the emergency may require a more sophisticated unit than the one at the nearest center; or (3) the nature of the emergency may require a less sophisticated unit than the one at the nearest center and there is a simultaneous need for the sophisticated unit at another emergency.

Figure 8.1   **Facility locations $\beta = 0.025$**

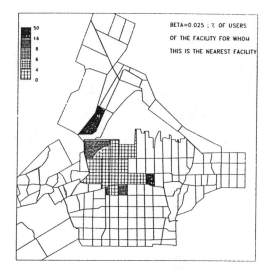

Figure 8.2   **Facility locations $\beta = 0.2$**

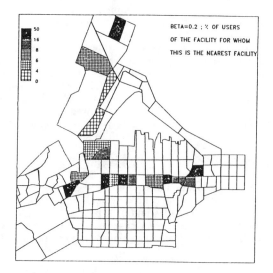

Table 8.1  **Location-Allocation, β=0.025**

| LOCATION | INFLOW | % NEAR | TIME |
|---|---|---|---|
| 906 | 0.05008 | 10.69 | 9.99130 |
| 1009 | 0.05000 | 2.49 | 10.05423 |
| 1010 | 0.04965 | 13.70 | 10.36131 |
| 101 | 0.04995 | 12.90 | 10.20260 |
| 102 | 0.04996 | 0.43 | 10.21568 |
| 103 | 0.04985 | 1.02 | 10.31112 |
| 201 | 0.04995 | 0.44 | 10.21159 |
| 202 | 0.04953 | 3.67 | 10.52506 |
| 704 | 0.04990 | 1.73 | 10.23997 |
| 604 | 0.04986 | 0.94 | 10.27149 |
| 603 | 0.04956 | 1.28 | 10.50371 |
| 508 | 0.04982 | 2.10 | 10.27036 |
| 703 | 0.05010 | 2.36 | 10.05465 |
| 601 | 0.05085 | 2.55 | 9.47751 |
| 602 | 0.05084 | 1.53 | 9.47516 |
| 501 | 0.05084 | 1.03 | 9.43295 |
| 502 | 0.05032 | 0.84 | 9.81265 |
| 503 | 0.04993 | 2.60 | 10.07693 |
| 504 | 0.04957 | 27.11 | 10.30252 |
| 1803 | 0.04943 | 24.80 | 10.61086 |

Table 8.2  **Location-Allocation, β = 0.20**

| LOCATION | INFLOW | % NEAR | TIME |
|---|---|---|---|
| 101 | 0.04998 | 5.76 | 8.33937 |
| 102 | 0.04842 | 11.71 | 8.43474 |
| 306 | 0.04167 | 30.19 | 7.10565 |
| 601 | 0.05910 | 22.63 | 7.33365 |
| 602 | 0.05791 | 5.94 | 7.24831 |
| 501 | 0.05981 | 26.50 | 6.91516 |
| 502 | 0.05480 | 12.32 | 7.17670 |
| 504 | 0.05134 | 6.95 | 7.18778 |
| 403 | 0.05125 | 7.35 | 7.00691 |
| 404 | 0.05091 | 25.68 | 6.85981 |
| 405 | 0.04933 | 9.62 | 6.81975 |
| 406 | 0.04830 | 14.18 | 6.74652 |
| 304 | 0.04437 | 9.17 | 6.95107 |
| 1803 | 0.04782 | 0.00 | 8.51377 |
| 1805 | 0.04821 | 2.59 | 8.15379 |
| 1808 | 0.04399 | 6.45 | 8.17632 |
| 1814 | 0.04988 | 16.34 | 7.56769 |
| 1905 | 0.04810 | 48.94 | 7.32244 |
| 804 | 0.04856 | 7.74 | 8.28779 |
| 803 | 0.04624 | 38.44 | 8.25054 |

Source: Tables based on O'Kelly (1987a).

Recent research research in the sphere of locational analysis has devoted considerable attention to the problem of backup systems in the event that the nearest unit is unavailable. A major review of these models appears in Daskin (1987) and in Hogan and ReVelle (1983).

Two implications emerge: first, there exists the possibility of a non-optimal response to the emergency in terms of travel time; and second, the notion that whenever there are two or more levels of service response (e.g., in EMS terminology, advanced life support (ALS) and basic life support (BLS)), it will often be necessary to use the most appropriate level of service, taking the current location of units into account. The thrust of the above discussion is that an emergency at a particular location may sometimes be answered by a unit other than the nearest. Everyday experience also suggests that multiple units often respond to an emergency to provide backup services. That is, situations may arise where a lower level (i.e., BLS) vehicle is dispatched in order to begin life saving procedures until the more appropriate higher level (i.e., ALS) vehicle is able to answer the call for service. The result of these observations is obvious - emergencies at certain locations are likely to produce responses from different facilities at different times. In this situation, a model of an EMS facility location pattern should incorporate a stochastic allocation mechanism, even in the case of a delivery system. Such a mechanism and its implied location of facilities is described in the next section.

## 8.8   A Probabilistic Location-Allocation Model for a Two Level EMS System

Suppose that two classes of facilities are to be located. Indicate their location patterns by $Y_{jh}$ ($h \in A,B$), where $Y_{jh} = 1$ if a type h facility is located in j and is equal to 0 otherwise. Further suppose that two distinct types of emergencies occur. Let type A emergencies require extremely sophisticated equipment (i.e. ALS) which because of cost considerations can only be provided from a relatively small number of sites. Suppose that the frequency of demands for ALS and BLS services at location i are known and are indicated by $O_{ih}$ ($h \in A,B$). Finally, suppose that a known impedance factor $\beta(h)$ characterises the responsiveness of facilities to distant calls. (This parameter describes the impeding effect of distance on the probability of a response to an emergency at i by a unit at j.) Let the travel time to the emergency site i from the facility at j be $c_{ij}$. Flows from facilities to emergencies are modeled as:

$$S_{ijh} = O_{ih} \, Y_{jh} \, \exp(-\beta(h) \, c_{ij}) / ( \sum_j Y_{jh} \, \exp(-\beta(h) \, c_{ij})) \tag{8.25}$$

where $S_{ijh}$ is the proportion of type h emergencies at i answered by a type h facility at location j. For a given set of facility locations, as the $\beta$ parameter is increased, the term

$$Y_{jh} \, \exp(-\beta(h) \, c_{ij}) / ( \sum_j Y_{jh} \, \exp(-\beta(h) \, c_{ij})) \tag{8.26}$$

tends to $X_{ijh}$, where $X_{ijh} = 1$ if the travel time from i to j is smaller than the travel time from i to any other type h facility; and $X_{ijh} = 0$ otherwise. That is, as $\beta(h)$ gets larger in absolute value, the proportion of emergency calls at i answered by the nearest unit tends to 1. As the distance impedance parameter gets larger the expression $S_{ijh}$ tends to $O_{ih} X_{ijh}$ and this model allocates the nearest unit to an emergency in the same fashion as the p-median problem. However when the parameter $\beta(h)$ is small (in absolute value) then the chances of a call for service at i being answered by the nearest facility is diminished.

Ignoring the issue of hierarchical level for a moment, perhaps some further interpretative comments are needed. The probabilistic allocation mechanism hinges on a spatial interaction model with a travel impedance parameter, $\beta$. The effect of introducing a probabilistic allocation mechanism varies with the value of $\beta$. For large values of $\beta$ (which enters the allocation model negatively) all calls for emergency services are answered by the nearest facility.  If two or more facilities are equally near to a demand node i, then each of these facilities respond to an equal proportion of the calls there.  At high values of $\beta$ an emergency at i is <u>rarely</u> answered by a facility other than the nearest one.  Intuitively, the value of $\beta$ reflects the degree to which the system is able to allocate the nearest unit to an emergency.  It is obvious that in the best of all possible worlds the nearest unit would always be dispatched to the emergency.  However, the fewer the units available at any particular level of the EMS system the greater are the chances that the nearest unit is unavailable.  Suppose that ALS units are so expensive to equip and maintain that they can be provided at a relatively small number of sites by comparison to the less expensive and more numerous BLS units.  In this model it makes sense to assume that the ALS system, with a small number of units, is represented by a probabilistic allocation mechanism with a small impedance parameter - because the chances of the nearest unit being unavailable are relatively large.  On the other hand, the BLS system is represented by a probabilistic allocation mechanism with a larger negative travel impedance parameter.  The resulting optimal location patterns can be shown to be more clustered for ALS units than for BLS units.  Numerical experiments show that as distance impedance decreases, the sum of the interfacility distances also decreases.  Intuitive arguments support this simple regularity - if only a few facilities can be located and if the chances of the nearest unit being unavailable are high, then the second nearest unit should not be located very far from the first.  The only way for this outcome to hold for all facilities is if they cluster together. Now if the second layer of facilities is superimposed on the first and if these units are more numerous, then their optimal location pattern is expected to be more decentralised than the higher order units.  Here it is assumed that ALS vehicles are backed-up by other units at the advanced support level - hence the clustered locational pattern; BLS services are to some extent complimentary to the ALS level.  To illustrate more concretely, we now examine a specific locational hierarchy.  Since the facility siting literature contains numerous examples of the more common deterministic service system, the hierarchy modelled below is that of a multilevel EMS organisation characterised by behavioural linkages, in a stochastic framework.

## 8.9  Results

In order to provide some evidence for the importance of linking mechanisms in the context of the probabilistic location-allocation problem, a series of numerical experiments has been carried out.  Table 8.3 shows the demand levels at a set of sixteen nodes together with the travel times in minutes between the nodes.  This hypothetical data set is similar to the one used in Ruefli and Storbeck (1984), and in O'Kelly and Storbeck (1984).

The consequences of the ALS siting decisions may be communicated to the BLS level in a variety of ways, of which five are considered here.  These linkages are listed below and are considered in the following sub-sections: (1) independent hierarchical levels; (2) mutually exclusive hierarchical levels; (3) successively inclusive hierarchical levels; (4) linked hierarchical levels; (5) linked levels - symmetric blocking.

Table 8.3  **Demand Levels and Inter-Nodal Travel Times**

| ID | 1 | 2 | 3 | 4 | 5 | 6 | 7 | 8 | 9 | 10 | 11 | 12 | 13 | 14 | 15 | 16 | $O_i$ |
|---|---|---|---|---|---|---|---|---|---|---|---|---|---|---|---|---|---|
| 1 | 0 | 3 | 7 | 15 | 11 | 8 | 4 | 11 | 16 | 21 | 16 | 13 | 16 | 17 | 19 | 24 | 0.14 |
| 2 | 3 | 0 | 6 | 12 | 8 | 5 | 5 | 8 | 13 | 8 | 13 | 10 | 13 | 18 | 16 | 21 | 0.10 |
| 3 | 7 | 6 | 0 | 8 | 4 | 7 | 11 | 10 | 9 | 14 | 11 | 16 | 17 | 24 | 16 | 17 | 0.07 |
| 4 | 15 | 12 | 8 | 0 | 4 | 7 | 11 | 10 | 7 | 8 | 11 | 16 | 17 | 24 | 16 | 11 | 0.03 |
| 5 | 11 | 8 | 4 | 4 | 0 | 3 | 7 | 6 | 5 | 10 | 7 | 12 | 12 | 20 | 12 | 13 | 0.05 |
| 6 | 8 | 5 | 7 | 7 | 3 | 0 | 4 | 3 | 8 | 13 | 8 | 9 | 10 | 17 | 11 | 16 | 0.11 |
| 7 | 4 | 5 | 11 | 11 | 7 | 4 | 0 | 7 | 12 | 17 | 12 | 9 | 12 | 13 | 15 | 20 | 0.17 |
| 8 | 11 | 8 | 10 | 10 | 6 | 3 | 7 | 0 | 5 | 10 | 5 | 6 | 7 | 14 | 8 | 13 | 0.05 |
| 9 | 16 | 13 | 9 | 7 | 5 | 8 | 12 | 5 | 0 | 5 | 4 | 9 | 19 | 17 | 9 | 8 | 0.03 |
| 10 | 21 | 18 | 14 | 8 | 10 | 13 | 17 | 10 | 5 | 0 | 5 | 8 | 9 | 16 | 8 | 2 | 0.02 |
| 11 | 16 | 13 | 11 | 11 | 7 | 8 | 12 | 5 | 4 | 5 | 0 | 5 | 6 | 13 | 5 | 8 | 0.03 |
| 12 | 13 | 10 | 16 | 16 | 12 | 9 | 9 | 6 | 9 | 8 | 5 | 0 | 3 | 8 | 6 | 11 | 0.03 |
| 13 | 16 | 13 | 17 | 17 | 13 | 10 | 12 | 7 | 10 | 9 | 6 | 3 | 0 | 7 | 3 | 8 | 0.02 |
| 14 | 17 | 18 | 24 | 24 | 20 | 17 | 13 | 14 | 17 | 16 | 13 | 8 | 7 | 0 | 8 | 13 | 0.01 |
| 15 | 19 | 16 | 16 | 16 | 12 | 11 | 15 | 8 | 9 | 8 | 5 | 6 | 3 | 8 | 0 | 5 | 0.02 |
| 16 | 24 | 21 | 17 | 11 | 13 | 16 | 20 | 13 | 8 | 3 | 8 | 11 | 8 | 13 | 5 | 0 | 0.08 |

Scenario 1: Independent hierarchical levels.  In the first scenario ALS and BLS facilities are located independently, but it is assumed that the impedance parameter in the ALS model is smaller than the BLS impedance parameter reflecting the slightly lower probability of the nearest ALS unit being able to respond to emergencies.  Specifically, let $\beta(A)=0.2$ and $\beta(B)=0.3$ in the allocation equation.  Results from the dual solution procedure indicate optimal locations for ALS units at (6,7,11) while BLS units should be located at (6,7,16).  The allocations which exceed 30% of the demand at origin i are shown in Figures 8.3 and 8.4.  Notice that service areas overlap - for instance in the ALS case (Figure 8.3) demands at node 14 are served by centers 7 and 11, while in the BLS case (Figure 8.4) centers 16 and 6 share the demands at nodes 11 and 9.  In the probabilistic allocation model demand can be shared between equally distant units while in deterministic "all or nothing" allocation models ties are usually broken arbitrarily.  The average travel time required to service the system can be computed by taking the weighted sum of the distances separating the facilities from the demand nodes.  The value of average travel time for ALS units is 5.69 while the average travel time for BLS units is 4.6 minutes.  The lower average travel time for BLS units can be accounted for by the relatively small amount of dispersion in the allocation of units from these centers - in other words the BLS units are more likely to be dispatched to the demand nodes to which it is nearest, as would be expected from the larger impedance parameter.  Table 8.4 gives details of the solution characteristics of this system.

Table 8.4  **Results Scenarios 1 2 and 3**

|  | (h) | BKS | DUAL | ATT | p(h) |
|---|---|---|---|---|---|
| SCENARIO 1 |  |  |  |  |  |
| ALS  (6,7,11) | 0.2 | −16.4359 | −16.7689 | 5.69 | 3 |
| BLS  (6,7,16) | 0.3 | −9.3118 | −9.9182 | 4.60 | 3 |
| SCENARIO 2 |  |  |  |  |  |
| ALS  (6,7,11) | 0.2 | −16.4359 | −16.7689 | 5.69 | 3 |
| BLS  (1,5,16) | 0.3 | −8.9352 | −9.8104 | 4.56 | 3 |
| SCENARIO 3 |  |  |  |  |  |
| ALS  (6,7,11) | 0.2 | −16.4359 | −16.7689 | 5.69 | 3 |
| BLS  (1,2,6, 7,11,16) | 0.3 | −11.8326 | −12.2243 | 3.99 | 6 |

## NOTES

$\beta$(h) The travel impedance parameter at each level
BKS  Best known solution value for the primal objective
DUAL Lower bound on the primal objective
ATT  Average travel time
p(h) The number of facilities at each level

Figure 8.3  **Scenario 1, ALS, no constraints, $\beta$(A)=0.2, p(A)=3**

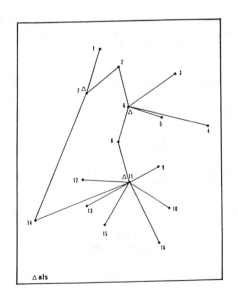

**Figure 8.4  Scenario 1, BLS, no constraints, β(B)=0.3, p(B)=3**

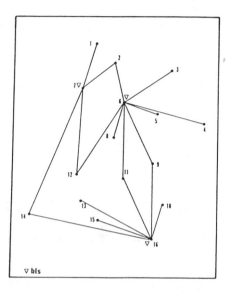

The major contrast between the ALS and the BLS location-allocation diagrams is the shift of the service area around facility 11 to a more decentralised service area around location 16. Notice what happens to node 12 - in the ALS system it was serviced by 11, but in the BLS system it is serviced by 7. Similarly, node 9 which is largely serviced by 11 in the ALS system is split between 16 and 6 in the BLS case. Ruefli and Storbeck (1984) report that for two service systems running in parallel, different location covering patterns are seen, but this was simply the result of different assumed demand surfaces for the two levels of service. Furthermore, their experiments involved four units at each hierarchical level and also they imposed a maximum travel time constraint of 4 minutes. The results in this chapter are quite different as they are not dependent on different demand surfaces and they do not adopt a "covering" objective.

Scenario 2: Mutually exclusive hierarchical levels. In the second scenario, β(A)=0.2 and β(B)=0.3 as before but the locations of the ALS units are excluded from the list of potential sites for BLS units. This linkage between hierarchical levels is achieved by conveying the results of the ALS siting to the BLS part of the program in the form of locational constraints. The resulting location-allocation pattern for BLS units is shown in Figure 8.5. The BLS units are located in (1,5,16). The value of average travel time for this locational pattern is 4.56 minutes - a slight decrease from the previous model. As might be expected, the primal objective function is less negative in this case since the use of locational constraints adds information to the program and hence reduces uncertainty.

Scenario 3: Successively inclusive hierarchical levels. In this analysis the ALS unit sites are explicitly assumed to provide BLS services as well as advanced service. This form of hierarchical organisation is termed "successively inclusive" since high order centers also provide lower order services. (The best-known hierarchical delivery system - Christaller's central place model - also makes this assumption.) Given that ALS units are located at (6,7,11) when β(A)=0.2, the optimal locations for BLS units are (1,2,6,7,11,16) with

$\beta(B)=0.3$ and 6 facilities. The allocations which exceed one sixth of the demand generated at node i are shown in Figure 8.6. (The lower cut-off level reflects the increased dispersion of allocations in a system with 6 facilities.) Table 8.4 summarises the solution characteristics : average travel time for BLS units is lower (just under 4 minutes) as is expected when extra facilities are located. Of course the primal objective is now more negative reflecting the increased uncertainty inherent in the allocation of 6 units. The ALS centers which respond to more than 30% of the service demand at node i can be deduced from Figure 8.3 while a list of the BLS units covering more than one sixth of the demand at i is easily found from Figure 8.6. It is readily determined that some facilities are more in demand than others. For example the ALS and BLS units at center 7 are the major units for only four demand nodes. On the other hand, the ALS unit in center 11 serves 9 nodes and the BLS unit there serves 7 nodes. Capacity constraints ought to be built in to future versions of the model.

Figure 8.5  **Scenario 2, BLS, mutually exclusive locations, $\beta(B)=0.2$, $p(B)=3$**

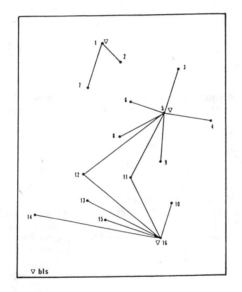

Figure 8.6 **Scenario 3, ALS & BLS, successively inclusive locations, β(A)=0.2, p(A)=3, β(B)=0.3, p(B)=6**

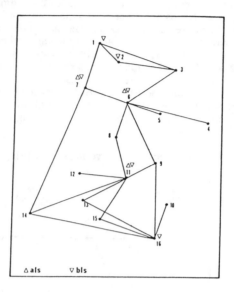

Scenario 4: Linked hierarchical levels. Suppose that three ALS units are located as in scenario 1, and that these centers have the following service patterns:

Center 6 serves (2,3,4,5,6,8)
Center 7 serves (1,2,7,14)
Center 11 serves (8,9,10,11,12,13,14,15,16)

Suppose now BLS unit locations are chosen to avoid duplication of these links while at the same time not excluding the possibility of a facility locating at (6,7,11). (That is while 6 cannot serve demand at 2,3,4,5,6 or 8 a unit located at 6 could still conceivably be able to service other demands.) This connection between hierarchical levels is achieved by running the model for the ALS units and then setting the elements of the travel time array equal to a large number whenever the corresponding entry in the ALS allocation table is greater than 30%. The effect of this device is to make it extremely costly to duplicate ALS allocations at the BLS level. This form of hierarchical linkage is a behavioural one in the sense defined above. Notice that the location of the BLS system is determined by the consequences of the siting of ALS centers: that is, the service areas adequately serviced by the ALS centers are eliminated from consideration for BLS sites. The results of this adjustment are shown in Figure 8.7. In that diagram ALS linkages are shown and it is clear that the BLS system does not reproduce these links. BLS units should be located in (1,5,16) and the characteristics of the system are summarised in Table 8.5.

Table 8.5  **Results Scenarios 4 and 5**

|  | (h) | BKS | DUAL | ATT | p(h) |
|---|---|---|---|---|---|
| SCENARIO 4 |  |  |  |  |  |
| ALS (6,7,11) | 0.2 | −16.4359 | −16.7689 | 5.69 | 3 |
| BLS (1,5,16) | 0.3 | −8.9352 | −9.8106 | 4.56 | 3 |
| SCENARIO 5 |  |  |  |  |  |
| ALS (6,7,11) | 0.2 | −16.4359 | −16.7689 | 5.69 | 3 |
| BLS (1,5,12) | 0.3 | −7.7900 | −8.4687 | 5.84 | 3 |

NOTES

$\beta$(h) The travel impedance parameter at each level
BKS  Best known solution value for the primal objective
DUAL Lower bound on the primal objective
ATT  Average travel time
p(h) The number of facilities at each level

Figure 8.7  **Scenario 4: Behaviourally Linked Levels**

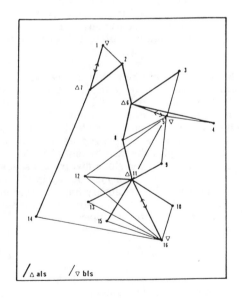

Figure 8.8   **Scenario 5: Behaviourally Linked Levels**

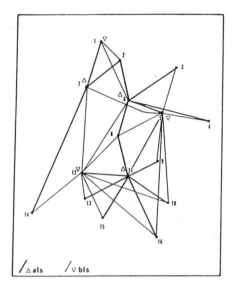

Obviously a comparison of the results of scenario 4 and scenario 2 indicate identical outcomes.   While the problem specification in scenario 4 does not exclude units from (6,7,11), it is extremely expensive for these centers to appear in the solution.   Thus the result arrived at through the technological-linkage of the hierarchical levels can also be generated without recourse to locational constraints.     Finally it remains to construct a behaviourally linked model which produces a different result from the technological-linkage case.

Scenario 5: Linked levels - symmetric blocking.   In the previous example information about the ALS allocation pattern was conveyed to the BLS system by an adjustment to the travel time array and this made it difficult for the BLS system to reproduce the ALS services.   However, there is nothing to prevent a BLS unit from serving ALS centers.   Thus there exists the strong possibility that an ALS center provides its services to a node which in turn provides it with BLS services.   (See for example the pairs (1,7), (5,6), and (11,16) in Figure 8.7.)   This outcome can be modified if the fact that an ALS unit services a node is used to block out BLS services in either direction along that particular link.   This model results are shown in Table 8.5 and Figure 8.8.   BLS units locate in (1,5,12) and the associated average travel time is 5.84 minutes.   The relatively long average travel time is expected since the ALS units have prevented the BLS units from using the shorter links.   While the average travel time is longer than in previous examples the primal objective is less negative - reflecting the lower dispersion of BLS services over the longer links which remain open.

8.10   Conclusions

The need for differentiating between technologically-linked and behaviourally-linked location hierarchies was demonstrated by examining a number of locational planning scenarios.   In certain instances, the solution of a siting problem at one level of a service hierarchy is necessarily dependent upon the actual siting decisions or siting requirements of another service level.   Two example situations which are characterised by such technological interdependencies are mutually exclusive and successively inclusive problems. These scenarios demonstrate the need for communicating the siting decisions made at one level to the locational process of another level.   In still other planning situations, the solution of the siting problem at one level of a service hierarchy may be dependent upon the service characteristics at another level.   Such linkages are appropriate for hierarchies offering services at different levels which are, in a sense, substitutable.   A common planning scenario which exhibits such behavioural interdependence is the siting of EMS systems which offer multilevel services and seek "back-up" coordination between levels. In this case, the situation dictates the communication of the deviations from siting objectives at one level to the locational process of another level, (Ruefli and Storbeck, 1984).   This chapter has suggested two new types of data which ought to be incorporated into models of hierarchical service delivery systems.   The first of these ideas is that different travel impedance parameters probably characterise the levels of a multi-level system.   The second idea is to link the location patterns of the levels of the system by constraints which are behavioural rather than technological - in brief this means that instead of conveying information about ALS locations to the BLS siting problem through locational constraints, we should consider instead information about the extent to which ALS services actually service the demand points.   Thus, in our small empirical example, a plausible scenario involved a constraint on the BLS system which prevented it from duplicating those services which were adequately provided by the ALS system.   The primary consequence of this development is the extension of a hierarchical linkage typology from simple deterministic models to some probabilistic situations.

# CHAPTER 9

## APPLICATIONS TO NETWORK DESIGN

### 9.1 Introduction

Where should facilities such as air terminals be located so as to connect the cities in a network? This question has contemporary relevance because all major air carriers in the U.S. operate a highly simplified sparse network organised around a "hub" in an attempt to minimise costs. Hubs are a type of facility located in a network in such a manner so as to provide a switching point for flows between other interacting nodes. These hubs are of great practical importance because they are at the heart of many express delivery systems, passenger networks, truck distribution networks, and communications systems. Siting a hub poses a problem for the spatial interaction theorist since the indirect routing of flows through central facilities implies a distortion of conventional transportation costs, which may require a re-evaluation of the demand for interaction between any two nodes. Even if the demand for interaction is inelastic (i.e. fixed with respect to network design), it is important to know just which flows are influential in the network's design.

The research reported here attempts to add to the literature dealing with interaction between hub facility location and spatial flows. Three types of situation are analyzed: (1) siting a single facility to service the interactions between a fixed set of nodes; (2) siting two (or more) planar facilities to service the interactions; (3) siting p hubs at discrete positions in a network to serve as switching points. These models are illustrated using intercity interactions from the U.S., and other applications to express package delivery and communication systems are suggested.

Before describing the precise nature of the analyses to be performed, the present paragraph outlines some real world examples to motivate the discussion. The single hub case is exemplified by the Federal Express Corporation network which has its hub in Memphis, Tennessee. This type of system requires virtually all interactions between pairs of cities on their network to be channelled through the hub. The advantage from the operator's point of view is the extremely sparse network needed to connect places. A similar arrangement is employed by several express package delivery firms (EPDFs). Chan and Ponder (1979) report that Federal Express used regional sorting centers in Pittsburgh and Salt Lake City as well as the major hub at Memphis in the early days of their operation. Increased aircraft range and capacity expansions at the Memphis hub have now produced a single facility system. Their network is being continually refined and updated; recent modifications include regional sorting centers in Oakland, California, and Newark, New Jersey. The need for careful siting of the hub facility is borne out by a consideration of the nature of flows of people and information. Air passenger hubs must be sited so as to connect high volume interactions with the minimum of inconvenience to passengers - otherwise a competitor with better service could lure away customers. Similarly EPDFs have greater ability to provide overnight service according to their strategic choice of site.

Other examples of application of hub location are suggested by the organisation of telephone networks (see Bell, 1977, page 86) and hub networks servicing the flow of information between computers on a university campus. A similar strategy is employed by Rockwell International, in its interplant communications system (see Figure 9.1).
A system, like the one shown in Figure 9.1, which is organised around more than one hub, is more difficult to plan than a single hub network because it is necessary to represent both the choice of facility location as well as the routing of interactions through the resulting network. This routing is primarily determined by considerations of distance and cost.

Schwartz (1987, Chapter 6) applies some shortest path algorithms to the problem; however for telecommunications through a network additional factors including capacity, redundancy and backup are often noted.

Figure 9.1  **Example: 4-hubs for Rockwell's interplant communications system**

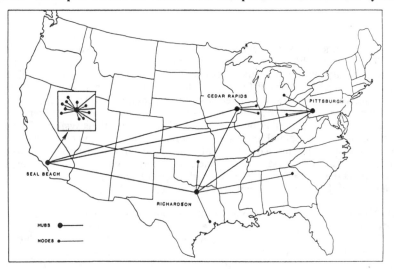

Source: Looney (1987) *Telecommunications* vol. 21(12), page 62.

The aim of the chapter is to provide simple models for the location of hub facilities. The single hub problem is easily dispensed with in both continuous and discrete form in the next section. Part 2 of the chapter describes a continuous two center version of the model, and solves versions with both exogenous and endogenous spatial interaction. Part 3 of the chapter analyses general multiple hub networks.

Throughout the chapter numerical results are reported using a 25 city data set representing a large proportion of the intercity passenger flows in the U.S. in 1970. (These data were analyzed in previous research by Fotheringham and Williams, 1983.) The data for this study are based on the airline passenger interactions between 25 U.S. cities in 1970 as evaluated by the Civil Aeronautics Board. The 25 cities which are listed below are part of a data set recording all flows between 100 cities. The cities included in this study are the larger places in the U.S. urban system and they account for 51% of all the flows observed. Cities in the interaction system are (1) Atlanta, (2) Baltimore, (3) Boston, (4) Chicago, (5) Cincinnati, (6) Cleveland, (7) Dallas-Ft Worth, (8) Denver, (9) Detroit, (10) Houston, (11) Kansas City, (12) Los Angeles, (13) Memphis, (14) Miami, (15) Minneapolis, (16) New Orleans, (17) New York, (18) Philadelphia, (19) Phoenix, (20) Pittsburgh, (21) St. Louis, (22) San Francisco, (23) Seattle, (24) Tampa, and (25) Washington D.C..

The models tackled in the rest of the chapter connect a facility location problem to a model of spatial interaction. It is recognised in this research that the hub facilities provide the focal points of a network through which intercity flows become articulated. This is not a new concern in transportation; after all the classic paper by Taaffe, Morrill and Gould (1963) emphasises the role of hubs and corridors in the development of the spatial economy. This research departs from the earlier efforts in several important respects: first, it describes a normative framework which finds the "best" location for hubs; second, the flows between hubs in the multi-hub model are an endogenous function of their relative location; and finally, a spatial interaction model is linked into the locational objective, thereby allowing for complete interdependence between interaction and facility location.

## 9.2 A Model of a Single Hub System

### 9.2.1 A Minisum Single Hub Location Problem

Suppose that flows between pairs of cities are known; denote these flows by $T_{ij}$, i,j=1,...,n. (In the examples that follow the interaction levels are scaled to sum to one, and $T_{ii} = 0$.) Consider the problem of choosing the location of a single hub facility such that all interactions flow through a single hub. The hub therefore acts as a clearing house or a switching point for all interactions. The location of the hub is denoted by Q=(X,Y). The locations of the n origins and destinations are denoted by $p_i = (x_i,y_i)$ for all i=1,...,n. Assume that every origin is also a destination so that the internodal distance matrix is square. The objective is to

$$\text{MIN}_{(Q)} \quad \Sigma_i\Sigma_j \, T_{ij}[d(p_i,Q) + d(Q,p_j)] \,, \tag{9.1}$$

where d(x,y) is the distance (in miles) between x and y. Denote by $O_i = \Sigma_j \, T_{ij}$ the total outflow from origin i, and by $D_j = \Sigma_i \, T_{ij}$ the total inflow to destination j. The problem may now be further simplified as :

$$\text{MIN}_{(Q)} \quad \Sigma_i \, d(p_i,Q) \, O_i + \Sigma_j \, d(p_j,Q) \, D_j \,, \tag{9.2}$$

which can easily be seen to be a classical Weber least cost location problem:

$$\text{MIN}_{(Q)} \quad \Sigma_k \, M_k \, d(p_k,Q) \,, \tag{9.3}$$

where $M_k = O_k + D_k$ is the total share of location k in all flows. Solution to this problem using a simple euclidean metric and Ostresh's (1973) program, results in a least cost site northeast of Cincinnati, Ohio. The objective function value at the optimum is 1481.72. If a site must be selected from among the 25 nodes of the system then Cincinnati (1490.4) is the cheapest location followed by Cleveland, Ohio. As is common in Weber models, the objective function is relatively flat in the vicinity of the solution; there are several cities with total cost values no more than 5% greater than the optimum: Pittsburgh (1511.6, +2%); Detroit (1523.4, +3%); and Chicago (1536.8, +4%). These results are consistent with previous studies which have shown that the area surrounding southern Ohio is efficient in providing access to the national market, (Harris, 1954). Although the empirical results are derived from passenger interaction data, there are some similarities to the observed siting of freight hubs by Emery (Dayton, Ohio); Airborne (Wilmington, Ohio); and DHL (Cincinnati, Ohio).

In the absence of scale effects and ignoring costs of setting up each intercity route, there is no rational reason to construct a hub oriented system, since by the triangle inequality:

$$\Sigma_i\Sigma_j \, T_{ij}d(p_i,p_j) \leq \Sigma_i\Sigma_j \, T_{ij}[d(p_i,Q) + d(p_j,Q)] \,. \tag{9.4}$$

Assuming however that there exist costs k associated with each intercity route, then a comparison of the one-hub system to a system without any hub indicates that a hub is preferable if :

$$\sum_i \sum_j T_{ij}[d(p_i,Q) + d(p_j,Q)] + kn$$

$$< \sum_i \sum_j T_{ij}d(p_i,p_j) + (k/2) \ n(n-1) \tag{9.5}$$

that is, if

$$\sum_i \sum_j T_{ij}[d(p_i,Q) + d(p_j,Q) - d(p_i,p_j)] \ < \ (k/2)(n^2 -3n) \tag{9.6}$$

or, in words, the incremental transportation costs must be less than the savings in link costs in order for a single hub to emerge. In the present case

$$\sum_i \sum_j T_{ij}d(p_i,p_j) = 921.91 \quad \text{and}$$

$$\sum_i \sum_j T_{ij}[d(p_i,Q) + d(p_j,Q)] = 1481.72. \tag{9.7}$$

These calculations show average interaction distances in miles. ($d(x,y)$ is in miles and the interaction data are scaled to sum to one.)   Total interaction costs could be obtained by multiplying by total volume of flow and a factor to represent costs per unit of volume per mile. The one-hub system is approximately 61% more expensive in terms of transportation costs. The decision to open a single hub to serve 25 cities would be based on an assessment of the savings from operating 25 routes instead of 300, in comparison to the increment to transportation costs.

### 9.2.2  A Minimax Single Hub Location Problem

We now extend the earlier research by formulating and solving a minimax hub location problem. The minimax hub location problem arises when the cost of the most costly interaction is required to be as cheap as possible. The rationale for the problem arises naturally from consideration of the indirect routing of flows via the hub facility. Obviously, some interacting nodes are poorly served by the hub network arrangement and it is the purpose of the minimax hub location model to make the most expensive interaction (in terms of travel time or distance) as small as possible. Existing geometric approaches to the minimax planar single facility location problem can be adapted this case. Consider the problem of choosing the location for a single hub facility such that the most costly flow (over all i and j) is as cheap as possible, given that the flows between i and j are routed from i to the hub and then from the hub to j. Mathematically, the objective is to choose the facility location so as to:

$$\begin{array}{ll} \text{MIN } G(Q) = & \text{MAX} \quad T_{ij}[d(p_i,Q) + d(Q,p_j)] \\ (Q) & (i,j) \end{array} \tag{9.8}$$

The interpretation of this objective requires some further elaboration. At a fixed hub location it is easy to evaluate the interaction costs of all the flows. These costs are simply the product of the volume of flow times the distance, via the hub, between the interacting nodes. The interacting pair with the largest cost is noted, and this quantity forms an upper bound on the solution. The objective is to move the hub to a site which makes this maximum cost as small as possible. It is important to note that the objective is not concerned with an aggregate cost of transportation, only with the most expensive of these interactions. The current problem contrasts with the previous model which places a hub so as to minimise the sum of interaction costs (see O'Kelly, 1986a). The minimum sum problem has the advantage of a separable objective function, which can be analyzed using classical optimisation techniques. The minimax objective poses a new problem compared with the minisum case: the minimax objective is not separable into inflow and outflow components since it is the combined cost of getting to and from the hub which must be considered. The general strategy in solving minimax location problems has been to use geometric reasoning to delimit the set of points in the plane which can service the associated nodes for no more than a fixed level of expenditure (i.e. a budget). This idea is applied in the case of the interaction between pairs of nodes by defining a disc which encloses all those points in the plane which can service the interaction for less than some fixed cost. There is one such disc for every interacting pair. Since only the larger of the two flows between every pair of nodes needs to be considered, a system with n nodes implies the construction and examination of n(n-1)/2 discs. Note that these discs are constructed on the basis of a fixed budget. At a very generous level of this budget, it is clear that the discs have large areas, with many points in common to all of them. By tightening the budget to a level such that the intersection of all the discs is reduced to a singleton (or to an arbitrarily small area) then the minimax location is found. The reason that this is the minimax location is that by definition only points inside the discs can service the interactions for less than the budget, and if there is only one point in common to all the discs, that point must be able to service all the interactions for no more than the budgeted amount. Any smaller budget would mean that some interacting pairs could not be serviced while any larger budget would be needlessly extravagant.

Suppose that an initial estimate of the most costly route is found for some proposed hub site, Q'. Call this initial estimate G(Q'). (This value is expressed in flow units x distance; e.g. passenger-miles.) The value G(Q') forms an upper bound on the objective. Initialise a lower bound on the objective as G(L), and a trial value of the objective is G* = (G(Q') + G(L))/2. For the expenditure of G* cost units, what is the set of feasible hub sites for an arbitrary interacting pair i and j ? The points in the plane satisfying this condition is defined to be the set:

$$D_{ij} = \{ \ (X,Y) \mid d(p_i,Q) + d(Q,p_j) \leq G^*/T_{ij} \ \}  \tag{9.9}$$

Refer to the points which satisfy (9.9) as a disc; the specific geometric properties depend on the measure $d(x,y)$ and will be given explicitly in the sequel. Construct an appropriate disc for every (i,j) pair. If there is a region which is common to all such discs it is clear that points within this common area can service all interactions for an expenditure of no more than G*. If G* is then varied until the intersection of the discs is as small as possible, a minimax location is obtained. (See Brady and Rosenthal, 1980; and Elzinga and Hearn, 1972.) In conventional minimax location problems the discs are either squares (rectilinear case) or circles (Euclidean case). Peeters (1980) provides very efficient special purpose algorithms for determination of the intersection question in these cases. His algorithms are implemented in routines reported in Hansen, Peeters, Richard and Thisse (1985) where constrained minimax problems with 200 nodes are solved in remarkably fast times. However, the hub minimax problem tackled here presents a different geometric problem: the appropriate discs are generally eight sided polygons (rectilinear case) or ellipses (Euclidean case).

A very powerful set of solution procedures is available for this problem once it is realised that the minimax hub location task is similar to the well known "round trip location problem" (Chan and Hearn, 1977; and the solution methods in Drezner, 1982). The round trip location model requires the siting of a fixed facility so that the most costly travel from an origin node to the facility, and from the facility to the destination, and then from the destination back to the origin (to form an obvious round trip) is as small as possible. Since the return from the destination to the origin introduces a constant quantity (the interpoint distance) which does not influence the solution, the minimax location model can be seen to be the same as a round trip problem with the modification that the interpoint distances are ignored. Extremely efficient procedures for this problem can therefore be based on the work in Chan and Hearn (1977) and Drezner (1982).

### 9.3   A Two-Hub Model in the Plane

It is difficult to service the entire continental U.S. from a single centrally located facility. This difficulty stems from the indirect routing of flows through the central facility. For passenger transportation services the savings accrued by minimising the number of routes are eventually out-weighed by the unwillingness of customers to make massive detours to get to their destinations. Thus it is unlikely that a traveller from Chicago to St. Louis would be willing to go via Pittsburgh as would be necessary on US Air.

This chapter continues with a discussion of the two-hub location problem. The route system in such a network is quite sparse as each city is linked to only one of the two hubs (an example is shown below in Figure 9.2). A crucial organisational feature is the routing of flows in one of two different ways: (i) if the origin-destination pair is linked to the same hub then interactions between them simply flow from the origin to the hub and then on to the destination. (ii) If, however, the origin and destination are linked to different hubs then flows must pass over the linkage between the facilities. It is in this latter case that the possibility for some unexplored transportational efficiencies occurs, for it is highly likely that the interaction flowing on the "link" between the hubs would enjoy substantial scale economies. The impacts of these economies on the spatial organisation of the system will be analyzed using the model developed for hub location.

The interactions between the origins and the facilities and between the facilities and the destinations are not without cost and this cost is assumed to be the straight line distance separating the nodes. Specifically, let **I** be the set of origin and destination indices and let **J** be the set of indices for hub facilities.

Figure 9.2 **Idealised Two Hub System**

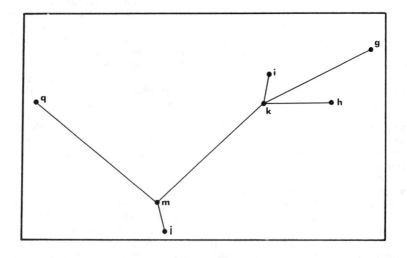

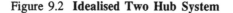

For all i in **I** and for all k in **J**, define the unit cost term to be:

$$d_{ik} = (((x_i - X_k)^2 + \sigma) + ((y_i - Y_k)^2 + \sigma))^{0.5} \qquad (9.10)$$

where $(x_i, y_i)$ is the pair of co-ordinates for node i, and $(X_k, Y_k)$ is the pair of co-ordinates for facility k. It is assumed therefore that distances are symmetric. The $\sigma$ term follows from a suggestion of Wesolowsky and Love (1972) and it guarantees that the distance function has continuous derivatives with respect to the hub locations. It is easy to show that the objective function evaluated with the sigma term adds a quantity proportional to the square root of $\sigma$ to the total. Thus, an answer arbitrarily close to the true optimum can be obtained by making $\sigma$ very small.

Interactions between facilities also involve a cost. The cost is assumed to be proportional to the straight line distance between the hubs multiplied by a constant proportionality factor. That is, for all j and k belonging to J, define

$$d_{jk} = \tau \left( ((X_j - X_k)^2 + \sigma) + ((Y_j - Y_k)^2 + \sigma) \right)^{0.5} \qquad (9.11)$$

where it is expected that a system displaying economies of scale in its linkage between facilities would have a proportionality factor of $\tau$ less than or equal to 1.

Let $u_i = 1$ if node i is serviced by hub 1, or 0 otherwise. Similarly, $v_i = 1$ if node i is serviced by hub 2 or 0 otherwise. The objective can now be formulated as :

$$\text{MIN} \quad \sum_i \sum_j T_{ij} \{ [u_i u_j (d_{i1} + d_{j1})] + [v_i v_j (d_{i2} + d_{j2})]$$
$$+ [u_i v_j (d_{i1} + d_{12} + d_{j2})] + [u_j v_i (d_{i2} + d_{12} + d_{j1})] \} . \qquad (9.12)$$

In (9.12) the minimisation is with respect to the locations of the two hubs. Only one of the products of the integer variables u and v is equal to one for any (i,j) pair. For a given set of integer variables (u,v) the problem can be solved by classical optimisation techniques as the objective is convex with continuous derivatives following the approach suggested by Wesolowsky and Love (1972). An illustration of the operation of the integer variables may be helpful here. For any pair of nodes i and j, the appropriate routing and transportation cost is indicated by the integer variables. Consider an eight node system where nodes 1,2,3,4,5 are connected to hub 1, while nodes 6,7 and 8 are connected to hub 2. Four types of connection are possible: (1) flows within the hub 1 service area (e.g. interactions between 1 and 3); (2) flows from the service area of hub 1 to hub 2 (e.g. interactions between 2 and 7); (3) flows from the service area of hub 2 to hub 1 (e.g. interactions between 8 and 3); and (4) flows within the hub 2 service area (e.g. interactions between 6 and 7).

The resulting interaction costs for each of the connections are described in the following notation; (for consistency with the distance definitions above the i and j subscripts refer to the interacting nodes, while the numerical subscripts are the two hubs):

TYPE 1: distance $= d_{i1} + d_{j1}$

TYPE 2: distance $= d_{i1} + d_{12} + d_{j2}$

TYPE 3: distance $= d_{i2} + d_{12} + d_{j1}$

TYPE 4: distance $= d_{i2} + d_{j2}$

The following notation represents a shorthand method of accounting for the various routing and transportation costs:

| k | $K_{ijk}$ | $d_{ijk}$ |
|---|-----------|-----------|
| 1 | $u_i u_j$ | $(d_{i1} + d_{j1})$ |
| 2 | $u_i v_j$ | $(d_{i1} + d_{12} + d_{j2})$ |
| 3 | $u_j v_i$ | $(d_{i2} + d_{12} + d_{j1})$ |
| 4 | $v_i v_j$ | $(d_{i2} + d_{j2})$ |

The objective can now be re-written:

$$\text{MIN} \quad T = \sum_i \sum_j T_{ij} \sum_k K_{ijk} \, d_{ijk} \, , \quad i,j=1,...,n; \; k=1,2,3,4. \qquad (9.13)$$
$$(Q)$$

where $Q=(X_1,Y_1,X_2,Y_2)$. For given values of the integer variables the solution procedure is to set the following partial derivatives to zero simultaneously:

$$\partial T / \partial Q_m = \sum_i \sum_j T_{ij} \sum_k K_{ijk} \, (\partial d_{ijk} / \partial Q_m) = 0, \quad m=1,...,4. \qquad (9.14)$$

The resulting system of four simultaneous equations can be solved by numerical methods. The analysis in the following paragraphs is limited by one major simplifying assumption: the assignment of demand nodes is fixed by creating non-overlapping partitions of the region. There are at most $n(n-1)/2$ such distinguishable cases (Ostresh, 1975). Facilities are then located on the basis of fixed $(u,v)$ values. That is, facility locations are chosen given that the nodes are assigned to them on the basis of the integer variables $(u,v)$. It remains to generate all feasible sets of integer variables. The method relies on the generation of all feasible non-overlapping partitions. This is accomplished using a suggestion of Ostresh (1975) for the two facility location-allocation problem. To partition the demand nodes into two sets, systematically cut the region in two. For each pair of points the integer variables $(u,v)$ are easily evaluated. In general there are $n(n-1)/2$ such patterns to be evaluated.

Some specifics are necessary here to explain exactly how the algorithm was implemented. The numerical procedure used to solve the system of equations is due to Powell (1970). The initial estimate of the solution borrows from the method used to start the well-known Weiszfeld procedure (Ostresh, 1978; Hansen, Peeters and Thisse, 1983) and are simply the weighted centroids of the nodes in each subset of the region. (Details of implementation and restrictions applying to the computer program which solves this model are presented in O'Kelly (1986a).) Using this starting method Powell's method solved the 25 node problem in 3 minutes 48 seconds on an IBM 3081.

As an application of the model, the parameter $\tau$ (which appears in the definition of the interfacility transportation costs) is systematically varied from 1 to 0.1 in order to determine the effect of distorted spatial impedance on the operation of the facilities. The results are summarised in Figure 9.3. As the cost of interfacility interaction declines the locations of the hubs spread apart, thereby increasing the amount of the system's costs allocated to the link. The aggregate system costs decline as the discount on the flow between the hubs increases. When $\tau=1.0$ there is no discount and the costs of operating the two hub network are 1346.70; as $\tau$ drops to 0.5 the costs drop to 1142.78; finally when costs of transportation on the link between the hubs are only one-tenth of the other interaction costs, the objective function value falls to 948.96. Notice the solution in Figure 9.3 for $\tau=1.0$. The cities of Dallas-Fort Worth and Houston are assigned to the eastern hub even though the optimally located western hub is closer. However, the complete enumeration of non-overlapping partitions also examined the western grouping of Dallas-Fort Worth and Houston. That this latter partition did not turn out to have a lower objective function value is an indication of the relative strength of flows from the Texan cities to the east coast. If a comparison is made to the one-hub system, it is clear that even in the undiscounted network the total interaction costs are smaller in the two hub system. It remains to establish the analytical conditions for a two rather than a one-hub system. Some connections in the two-hub model are cheaper while others are more expensive. The top panel of Figure 9.4 illustrates the type of linkages which experience cheaper transport costs in the two-hub system. It is easy to show that the distance between i and j via H1 is less than the distance through H. Similarly k to l via H2 is shorter than k to l via H. The lower panel of Figure 9.4 shows that the linkage between i and k is longer using the dotted route (2 hubs) than the solid route (1 hub). In aggregate the two-hub system is cheaper if the following condition holds:

$$\sum_i \sum_j T_{ij} \{ \ u_i u_j \ [d(p_i,Q) + d(p_j,Q) - d_{i1} - d_{j1}]$$

$$+ \ v_i v_j \ [d(p_i,Q) + d(p_j,Q) - d_{i2} - d_{j2}] \ \} \ >$$

$$\sum_i \sum_j T_{ij} \{ \ u_i v_j \ [d_{i1} + d_{12} + d_{j2} - d(p_i,Q) - d(p_j,Q)] \tag{9.15}$$

$$+ \ u_j v_i \ [d_{i2} + d_{12} + d_{j1} - d(p_i,Q) - d(p_j,Q)] \ \} \ + Z$$

where Z is the cost of the interfacility linkage. Note that the two-hub model has one extra interfacility connection than the one-hub case. This inequality holds whenever the savings due to local handling of the cities in the service areas of the two hubs outweighs the extra costs of routing the flows between regions over the relatively longer link. If this condition does not hold then there is no rationale for a two-hub system. If however the costs are lower in the two-hub system, then the incentives are towards regional processing. The inequality is more likely to hold when the scale effects in transportation over the linkage are strong. It is clear from the data above that the single hub system is always more expensive in our present case, unless the cost of constructing the link between the two hubs is extremely high relative to the costs of constructing the other links. In the next two sections two possible extensions of the model are suggested: (1) the development of notation for a three-hub planar model; and (2) the replacement of the exogenous spatial flows between the nodes with terms which vary with the relative costs of routing the flow through the hub oriented network.

Figure 9.3  **Results: Two Hub System with τ=1.0**

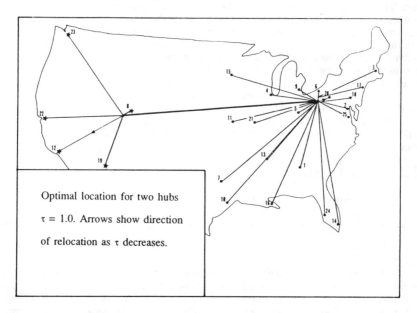

Optimal location for two hubs

τ = 1.0. Arrows show direction

of relocation as τ decreases.

Figure 9.4  **Comparison of Shorter and Longer Routes in a Two Hub System**

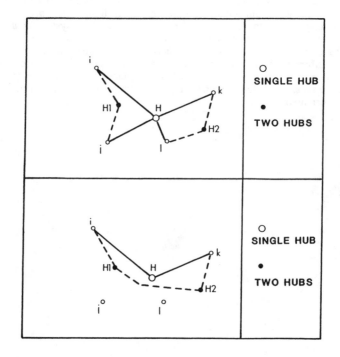

## 9.4   Notation Modification for a Three-Hub System

In order to show that the model is not necessarily restricted to the two-hub situation, we now develop some notation for a three-hub planar model. We use the numerical subscript 1 2 or 3 to indicate the hub in question and maintain the convention that an i or j subscript refers to an origin or destination node. Three sets of integer indicators are needed: $u_i$, $v_i$, and $w_i$, which have the following definitions:

$u_i$ = 1 if node i is connected to hub 1, and is zero otherwise,
$v_i$ = 1 if node i is connected to hub 2, and is zero otherwise,
$w_i$ = 1 if node i is connected to hub 3, and is zero otherwise.

Then the following are the nine possible types of routing:

| k | $K_{ijk}$ | $d_{ijk}$ |
|---|---|---|
| 1 | $u_i u_j$ | $(d_{i1} + d_{j1})$ |
| 2 | $u_i v_j$ | $(d_{i1} + d_{12} + d_{j2})$ |
| 3 | $u_i w_j$ | $(d_{i1} + d_{13} + d_{j3})$ |
| 4 | $v_i u_j$ | $(d_{i2} + d_{12} + d_{j1})$ |
| 5 | $v_i v_j$ | $(d_{i2} + d_{j2})$ |
| 6 | $v_i w_j$ | $(d_{i2} + d_{23} + d_{j3})$ |
| 7 | $w_i u_j$ | $(d_{i3} + d_{13} + d_{j1})$ |
| 8 | $w_i v_j$ | $(d_{i3} + d_{23} + d_{j2})$ |
| 9 | $w_i w_j$ | $(d_{i3} + d_{j3})$ |

The objective function is exactly as before except that the summation is over nine different routing types (as opposed to four above); and there are six choice variables (the 3 pairs of coordinates); that is the objective can be written:

$$\text{MIN } T = \sum_i \sum_j T_{ij} \sum_k K_{ijk} d_{ijk}, \quad i,j=1,...,n; \; k=1,...,9 , \qquad (9.16)$$
$$(Q)$$

where $Q=(X_1,Y_1,X_2,Y_2,X_3,Y_3)$. For given values of the integer variables the solution procedure is to set the following partial derivatives to zero simultaneously:

$$\partial T/\partial Q_m = \sum_i \sum_j T_{ij} \sum_k K_{ijk} (\partial d_{ijk}/\partial Q_m) = 0, \quad m=1,...,6. \qquad (9.17)$$

The resulting system of six simultaneous equations can be solved by numerical methods in the same fashion as the previous model. The generation of all possible non-overlapping partitions into three regions poses a considerable computational task (see also Ostresh, 1975, page 215), and the method would still lack the generality because of the non-overlapping assumption. Rather than pursue the further disaggregation of the planar hub model, these extensions are postponed until Part 3 of this chapter where a discrete version of the model is formulated. A more interesting extension is to pursue the idea of making the flows between the interacting nodes a function of their actual spatial separation when connected via a hub network.

## 9.5   Endogenous Spatial Interaction

Recall the definition of $K_{ijk}$ and $d_{ijk}$ in the previous section. It may well be that certain linkages (i,j) involve substantial interaction costs despite the careful choice of hub locations. In spatial interaction theory it is expected that flows between places vary inversely with their spatial separation. Thus, the internodal flows might be influenced by the organisation of the system of hub locations. To model this linkage assume that $T_{ij}$ may be represented by:

$$T_{ij} = O_i D_j \exp(-\beta(\Sigma_k K_{ijk} d_{ijk})) / \Sigma_q D_q \exp(-\beta(\Sigma_k K_{iqk} d_{iqk})) \tag{9.18}$$

where, as before, $\Sigma_k K_{ijk} d_{ijk}$ evaluates the appropriate routing and transportation costs. The revised optimisation problem can now be written as:

$$\underset{(W,Q)}{\text{MIN}} \quad T = \Sigma_i \Sigma_j \; T_{ij} \; [\Sigma_k K_{ijk} d_{ijk}] \tag{9.19}$$

where $T_{ij}$ is defined above. The optimisation problem can therefore be solved for given integer variables finding the coordinates of the hubs  to solve the following system of equations:

$$\partial T/\partial Q_m = \Sigma_i \Sigma_j \; (T_{ij} \; [\partial(\Sigma_k K_{ijk} d_{ijk})/\partial Q_m]) \tag{9.20}$$

$$+ \quad \Sigma_i \Sigma_j \; ([\Sigma_k K_{ijk} d_{ijk}][\partial T_{ij}/\partial Q_m]) = 0 \quad m = 1,2,3,4.$$

The major task is the calculation of $\partial T_{ij}/\partial Q_m$ which is reported in O'Kelly (1986a). These equations are more difficult to solve, and 15 - 20 minutes IBM 3081 CPU time was required to find facility locations.

Before explaining the results in some detail, the role of the parameter $\beta$, which has already played a crucial role in our earlier discussions, should be clarified. Note that $\beta$ governs the sensitivity of interactions to the relative distances to alternative destinations. If $\beta$ is large, then a disproportionate share of flows from origin i travel to the nearest destination. As $\beta$ tends to zero, flows become relatively insensitive to the various distances which must be travelled to reach different destinations. These comments simply repeat well-known properties of the production-constrained spatial interaction model (as described in Chapter 2). Notice now that alterations in hub location change the relative distances to destinations. Thus, a place which is closest to origin city i in a given hub system, could well be the furthest in another circumstance. The endogenous flows respond to the changing relative distances to destinations by adjusting towards  those alternatives which are closest. The magnitude of this adjustment depends on the value of the parameter $\beta$. At large values of $\beta$ major shifts in destination allocation can occur in response to changes in hub location. While it would be extremely difficult to estimate actual values of the parameter $\beta$, some simple sensitivity analysis can be performed to show how completely different spatial organisation can emerge from different parameter values. The data are now analyzed in a comprehensive manner to enable some understanding of the complex interrelationships which emerge when interaction and facility location are simultaneously determined. The measures to be reported below are explained here. Recall that the data used here are the actual intercity passenger flows from the Civil Aeronautics Board 1970 sample survey. Therefore there is a given set of flows against which to compare the model results. The actual data can be thought of as "exogenous" interaction data, while the results from the model in this section are "endogenously" calculated estimates of the same flows. It is clear that a comparison of the two flow matrices is needed. Measures of the volume of interaction are computed for the following subsets: interactions using hub 1 exclusively, (k=1); interactions using hub 2 exclusively, (k=2); interactions flowing from hub 1 to hub 2, (k=3); interactions flowing from hub 2 to hub 1, (k=4).

Define:
$$G_k = \sum_i \sum_j K_{ijk} T_{ij}$$

$$G^*_k = \sum_i \sum_j K_{ijk} T_{ij}^*$$

$$H_k = \sum_i \sum_j K_{ijk} \mid T_{ij} - T_{ij}^* \mid$$

where $G_k$ is the observed share of type k interaction, $G^*_k$ is the predicted share of type k interaction, $T_{ij}$ and $T_{ij}^*$ are the observed and predicted i to j flows, and $H_k$ is the sum of the absolute deviations between the observed and predicted flows of type k.

These quantities are now reported for a series of different parameter combinations.

### 9.5.1 Results for ($\beta$=0.0001, $\tau$=1.0), ($\beta$=0.0001, $\tau$=0.4)

The results of these computer runs are shown in Table 9.1. In this model, interactions are relatively stable with respect to changes in hub location because the propensity to allocate all the demand to the nearest destination is muted. The observed and predicted levels of flows and the levels of absolute deviation are shown in Table 9.1. In the case of $\tau$=1 (the left hand panel) the flows in the four components of the system are quite balanced. When the observed data are aggregated according to the two hubs shown in Figure 9.5A, 36.52%, 24.76%, 19.36% and 19.36% of the interactions occur in categories 1,2,3 and 4 respectively. (Since the observed interaction matrix is symmetric, it is easy to show that the flows between the two hubs are the same in both directions; in general the predicted interactions are not symmetric so that they can have unequal inter-facility flow.)  As can be seen from Table 9.1, the endogenously calculated interactions show smaller flows in categories 1 and 2 (30.18% and 18.01% respectively), and larger flows between the hubs (25.73% and 26.13% respectively in categories 3 and 4) than the observed data. Overall, the results in the $\tau$=1.0 case suggest that if the hubs are located as in Figure 9.5A, and if interactions are allowed to respond to internodal distances, then major adjustments in flow would take place. In the $\tau$=0.4 case (right hand panel) the eastern hub handles a disproportionate share of the interaction (see Figure 9.5B). Notice that this is also a characteristic of the actual interactions when they are aggregated in the manner illustrated in these diagrams.

Figure 9.5A **Interaction Results**

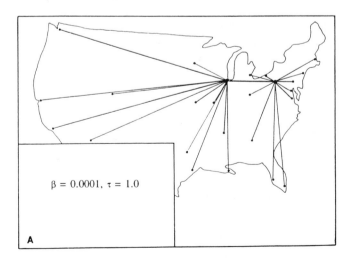

$\beta$ = 0.0001, $\tau$ = 1.0

A

Figure 9.5B  **Interaction Results**

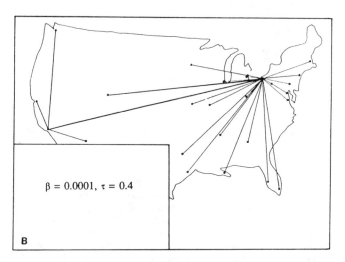

$\beta = 0.0001, \tau = 0.4$

B

Table 9.1  **Results β=0.0001, two values of τ**

(see notes at the end of Table 9.2)

|  | τ=1.0 | | | τ=0.4 | | |
|---|---|---|---|---|---|---|
|  | SUM | MIN | MAX | SUM | MIN | MAX |
| **Interaction serviced exclusively by hub 1** | | | | | | |
| TOBS | 0.3652 | 0.0001 | 0.0240 | 0.7332 | 0.0001 | 0.0240 |
| TPRED | 0.3018 | 0.0003 | 0.0128 | 0.7049 | 0.0002 | 0.0211 |
| DIFF | 0.1324 | 0.0000 | 0.0126 | 0.2841 | 0.0000 | 0.0129 |
| **Interaction serviced exclusively by hub 2** | | | | | | |
| TOBS | 0.2476 | 0.0001 | 0.0108 | 0.0484 | 0.0004 | 0.0108 |
| TPRED | 0.1801 | 0.0002 | 0.0085 | 0.0189 | 0.0003 | 0.0044 |
| DIFF | 0.1011 | 0.0000 | 0.0073 | 0.0295 | 0.0001 | 0.0064 |
| **Interaction serviced by hub 1 to 2** | | | | | | |
| TOBS | 0.1936 | 0.0001 | 0.0202 | 0.1092 | 0.0001 | 0.0124 |
| TPRED | 0.2573 | 0.0002 | 0.0213 | 0.1365 | 0.0002 | 0.0146 |
| DIFF | 0.0785 | 0.0000 | 0.0026 | 0.0423 | 0.0000 | 0.0024 |
| **Interaction serviced by hub 2 to 1** | | | | | | |
| TOBS | 0.1936 | 0.0001 | 0.0202 | 0.1092 | 0.0001 | 0.0124 |
| TPRED | 0.2613 | 0.0002 | 0.0192 | 0.1388 | 0.0002 | 0.0134 |
| DIFF | 0.0813 | 0.0000 | 0.0027 | 0.0434 | 0.0000 | 0.0023 |

Table 9.2  **Results β=0.001, two values of τ**

|        | τ=1.0 | | | τ=0.4 | | |
|--------|-------|-------|-------|-------|-------|-------|
|        | SUM   | MIN   | MAX   | SUM   | MIN   | MAX   |
| **Interaction serviced exclusively by hub 1** | | | | | | |
| TOBS   | 0.6506 | 0.0001 | 0.0240 | 0.1812 | 0.0001 | 0.0240 |
| TPRED  | 0.7660 | 0.0001 | 0.0274 | 0.2047 | 0.0006 | 0.0199 |
| DIFF   | 0.3050 | 0.0000 | 0.0112 | 0.0657 | 0.0000 | 0.0123 |
| **Interaction serviced exclusively by hub 2** | | | | | | |
| TOBS   | 0.0756 | 0.0002 | 0.0108 | 0.3842 | 0.0001 | 0.0108 |
| TPRED  | 0.1353 | 0.0005 | 0.0224 | 0.3112 | 0.0001 | 0.0125 |
| DIFF   | 0.0753 | 0.0001 | 0.0116 | 0.2146 | 0.0000 | 0.0098 |
| **Interaction serviced by hub 1 to 2** | | | | | | |
| TOBS   | 0.1369 | 0.0001 | 0.0124 | 0.2173 | 0.0001 | 0.0202 |
| TPRED  | 0.0205 | 0.0000 | 0.0027 | 0.1942 | 0.0001 | 0.0295 |
| DIFF   | 0.1164 | 0.0000 | 0.0097 | 0.1009 | 0.0000 | 0.0145 |
| **Interaction serviced by hub 2 to 1** | | | | | | |
| TOBS   | 0.1369 | 0.0001 | 0.0124 | 0.2173 | 0.0001 | 0.0202 |
| TPRED  | 0.0772 | 0.0000 | 0.0080 | 0.2904 | 0.0002 | 0.0262 |
| DIFF   | 0.0617 | 0.0000 | 0.0044 | 0.0961 | 0.0000 | 0.0074 |

NOTES TO TABLES 9.1 and 9.2

TOBS  = the observed flow from i to j
TPRED = the predicted flow from i to j
DIFF  = the absolute value of TOBS - TPRED
SUM   = the sum over the appropriate subset of the
        variable named in the row of the table
MIN   = the minimum value over the appropriate subset of the variable
        named in the row of the table
MAX   = the maximum value over the appropriate subset of the variable
        named in the row of the table.

Source: O'Kelly (1986a)

## 9.5.2   Results for ($\beta$=0.001, $\tau$=1.0), ($\beta$=0.001, $\tau$=0.4)

The results of these computer runs are shown in Table 9.2. When $\beta$=0.001 interactions are much more sensitive to facility locations. Any movement of the coordinates of a facility causes a recalculation of the interaction matrix. The solution of the systems of equations become more difficult to determine because of this increased level of interdependency of interaction and location. In the empirical results shown in Table 9.2 the case of $\tau$=1.0 shows very little use of the link between the two facilities, in fact the endogenous levels of interfacility interaction at these locations are especially small (2.05%) in the east to west direction compared to the actual levels of flow which the observed data suggest (13.69%). The data suggest that an interaction system with a large impedance parameter and without any scale effects on the interfacility linkage would essentially operate as two separate systems without much interconnection. A value of $\tau$=0.4 ameliorates the impact of the high $\beta$ parameter. High levels of interaction between the facilities emerge. The resulting network shows an interesting adjustment: the levels of interaction within the service area of hub 1 and from east to west are similar to the observed levels. (Observations in categories 1 and 3 are 18.12% and 21.73%; endogenous levels in the same categories are 20.47% and 19.42%.)  However, the flows within the service area of hub 2 decrease at the expense of flows from west to east, when the interactions are allowed to adjust to the facility locations.

## 9.6   Discrete Multi-Hub Model

The multi-hub location problem can be illustrated with a small example. Assume that n nodes interact in some way (e.g. passenger flows, express package deliveries or telecommunications). Each node is assumed to be  connected to one of p hubs, which act as the switching points. All the hubs are assumed to be interconnected. Thus, if i is connected to hub k and j is connected to hub m, flows from i to j are routed from i to k, from k to m, and from m to j. A simple example network was shown in Figure 9.2. The matrix of assignments will be explained fully in the notation section; for the present, note that the assignment of the nodes to hubs is denoted by integer variables $X_{ik}$. The transportation costs in a hub oriented network are critically dependent on the siting of the switching points. As will be shown, some of the components of these costs are computed from linear functions of the assignment variables  but more importantly there are also quadratic cost terms which arise from the inter-hub transactions. The first parts of this chapter dealt only with one and two hub systems in the plane. Now the model is extended by suggesting a programming formulation for the p-hub location problem. We now report a formulation of a general hub location model as a quadratic integer program. Computational results from two simple heuristics are presented for the task of siting 2, 3 or 4 hubs to serve interactions between sets of 10, 15, 20 and 25 U.S. cities.

## 9.7   A Quadratic Program for Interacting Facility Location

In the location model based on O'Kelly (1987b), the following notation is used: $X_{ik}$ = 1 if node i is linked to a hub at k, $X_{ik}$ = 0 if node i is not linked to a hub at k. The number of units of flow between nodes i and j (exogenous), is $T_{ij}$, and $T_{ii}$ = 0 by assumption. The transportation cost of a unit of flow between node i and j is $d_{ij}$; p is the total number of hubs to be constructed (exogenous), n is the total number of cities to be interconnected (exogenous). It is assumed without loss of generality that the flow matrix, W, is scaled so that all the elements sum to one. (In the examples discussed below the flow matrix is symmetric; the heuristics developed in this chapter do not rely on this symmetry. The cost matrix in the examples is also symmetric; again the heuristics do not rely on this.)  Note that the zero diagonal elements in the transportation cost matrix are required to ensure that no intrahub costs are assigned. The hub location problem can be written as :

$$\text{(QP) MIN}\quad Z = \sum_i \sum_j T_{ij} (\sum_k X_{ik} d_{ik} + \sum_m X_{jm} d_{jm} \tag{9.21}$$

$$+ \ \tau \ \Sigma_k \Sigma_m \ X_{ik} X_{jm} d_{km} \ )$$

S.T.   $(n-p+1) \ X_{jj} - \Sigma_i \ X_{ij} \geq 0$   for all j          (9.22)

$$\Sigma_j \ X_{ij} = 1 \qquad \text{for all i} \qquad (9.23)$$

$$\Sigma_j \ X_{jj} = p \qquad (9.24)$$

$X_{ij}$ is either 0 or 1.          (9.25)

In (9.21) the first two terms inside the brackets evaluate the cost of assigning a node to its hub for outgoing and incoming flows respectively. The second component (the double summation over k and m) counts the costs of those interactions which must flow between hubs. These inter-hub costs are multiplied by a parameter $\tau \leq 1$ to reflect the scale effects in interfacility flows. The idea is that the flows between hubs might enjoy a discounted transportation rate arising from the greater volume on these connections. For instance, a passenger airline might be able utilise a wide-body jet with a high load factor on a busy inter-hub linkage. In telecommunications a high capacity link between major switching centers would generate the same type of efficiency. The model is set up here to allow for experimentation with the costs of interfacility linkages versus the costs of the "spokes" on the hub network. Empirical analysis is needed to quantify the magnitude of actual scale effects on high volume network linkages. Constraint (9.22) ensures that no node is assigned to a location unless a hub is opened at that site. (That is, $X_{jj}$ must equal 1 before any other node can be allocated to j.)   Note that constraint (9.22) is written in the weak form following ReVelle and Swain (1970); the constraint works by recognising that nodes can be only be assigned to hubs and that at most $(n-p+1)$ nodes can be assigned to any hub (including the hub itself). Constraint (9.23) ensures that each node is assigned to one and only one hub (whenever X is integer). Constraint (9.24) generates the correct number of hubs. (The phrase "open facility" is used to mean a location where a hub is operating; for this location j, $X_{jj}=1$. Conversely a "closed facility" is a location where $X_{jj}=0$. The integer requirement on the allocation variables is crucial and is enforced by constraint (9.25).)   The objective can be reworked in the following manner:

$$\text{MIN} \ \ Z \ = \ \Sigma_i \Sigma_k \ X_{ik} d_{ik} \ [\Sigma_j \ T_{ij}] + \Sigma_j \Sigma_m \ X_{jm} d_{jm} \ [\Sigma_i \ T_{ij}] \qquad (9.26)$$

$$+ \ \tau \ \Sigma_j \Sigma_m \ X_{jm} \ \Sigma_i \Sigma_k \ X_{ik} T_{ij} d_{km}$$

Let $O_i = \Sigma_j \ T_{ij}$ be the total amount of interaction originating at node i; and similarly let $D_j = \Sigma_i \ T_{ij}$ be the total amount of flow having node j as its destination. The   objective can be re-written:

$$\text{MIN} \ \ Z \ = \ \Sigma_i \Sigma_k \ X_{ik} d_{ik} \ (O_i + D_i) \qquad (9.27)$$

$$+ \ \Sigma_j \Sigma_m \ X_{jm} \ \Sigma_i \Sigma_k \ X_{ik} (\tau \ T_{ij} d_{km}) \quad .$$

In (9.27) the first summation (over i and k) evaluates the cost of allocating  every node to its hub and weights this cost by the total volume of outbound and inbound traffic using this route. That is, multiply the transportation costs from node i to its associated hub by the volume of flow in both directions over this connection. By doing this for all cities in turn, we accumulate the portion of the total transportation costs associated with the "spokes" of the hub and spoke system. To understand the second term in (9.27), consider two nodes j and i. The innermost summation (over i and k) picks out the row of the cost matrix associated with the facility serving i. The second summation (over j and m) then picks out the element of this row associated with the facility serving j. The diagonal element is selected if both i and j are served by the same hub while in every other case the computation evaluates the cost of linking i and j through the hubs to which they are connected.

9.8  Related Research

The optimisation model described in section 9.7 is different from the p-median problem (PMP) and multifacility Weber problems (see Hansen, Peeters and Thisse, 1983, 237). In those models there is either no interaction between facilities (PMP) or else a fixed amount of flow between them (multifacility Weber problem). In the present model, the magnitude of the flow between the hubs, or the interfacility linkages, depends on the location of the hubs. Thus, the interfacility linkage costs are an endogenous function of the facility locations and this is a characteristic of a difficult class of location models (see Hansen, Peeters and Thisse 1983, 253). While the uncapacitated PMP clearly implies the assignment of nodes to their nearest facility at optimality, it is easy to show that the hub location problem cannot assume this allocation rule. To see this, consider node i in Figure 9.2. Suppose that i has strong interactions with nodes q, m and j, but that there are negligible interactions between i and g, k and h. It would make sense for i to be allocated to hub m rather than hub k since the bulk of i's interaction is with the nodes which are close to m. Since the allocation of a node to a hub is based on that hub's ability to service the interaction pattern, (i.e. a derived demand for connectivity), the proximity of the hub to the node itself may be unimportant.

Having made this point, it should now be stated that patterns of spatial interaction exhibit a strong distance decay effect, and so the hypothetical example suggested above (i interacting strongly with relatively remote nodes) is perhaps the exception rather than the rule. The implication from the preceding paragraph is this: the assignment of demand to hubs is not simply a by-product of the location of the hubs. Indeed, a generalised quadratic assignment model must be solved for each locational configuration of the hubs in order to find the exact solution to the full hub location problem. (Ross and Soland (1977) present a general class of assignment problems to solve location tasks; none of their models contain a quadratic term.)

The literature contains several  suggestions that a quadratic term evaluates the costs of interaction between facilities (see Hansen, 1979, 64). Gallo, Hammer and Simeone (1980) cite the potential application of the quadratic knapsack problem to the task of locating nodes in a communications network. The idea of land use plan design models using quadratic programming techniques was proposed by Brotchie (1969); similar ideas are explored in Lundqvist (1973) and Snickars (1978). Finally, Francis and White (1974) devote considerable attention to a quadratic assignment location problem, involving the location of n facilities among n potential sites where it is known that interfacility connections exist. Recently a heuristic procedure for this task has appeared (West, 1983). The model under consideration here is quite different from these existing references as the choice is that of p from n possible sites (p < n). Not only must the assignment of n nodes to p hubs be determined, but also the analysis must examine the choice of p facilities from n alternatives. In general the two tasks are not separable. By way of analogy, the hub location model stands in the same relationship to the quadratic assignment problem as does the PMP to the linear assignment problem (see Ross and Soland, 1977). The first step is the identification of a standard quadratic program from (QP) above. In what follows all vectors are column vectors; ' denotes transposition.

The objective can now be written:

$$\text{MIN} \quad c'x \; + \; x'Q \, x \tag{9.28}$$

S.T. $\mathbf{Ax} \geq \mathbf{b}$ where $X_{ij}$ is either 0 or 1. (9.29)

The matrix $\mathbf{A}$ and the vector $\mathbf{b}$ are easily developed from the constraints (9.22), (9.23) and (9.24). This program is convex if and only if $\mathbf{Q}$ is positive semi-definite (PSD). (For this discussion assume $\mathbf{T}$ and $\mathbf{D}$ are symmetric, where $\mathbf{T}$ is the matrix of observed internodal flows and $\mathbf{D}$ is the matrix containing $d_{ij}$ values.) It is known that a symmetric matrix is PSD if and only if all of its eigenvalues are positive. Furthermore since $\tau\mathbf{Q} = \mathbf{T} \times \tau\mathbf{D}$ (where x denotes the Kronecker product) the eigenvalues of $\tau\mathbf{Q}$ are simply found: $q_{ij} = t_i d_j$, where $t_i$ and $d_j$ are the eigenvalues of $\mathbf{T}$ and $\tau\mathbf{D}$ respectively, (see Bellman, 1970, p. 235). As Snickars (1978) shows, the structure of $\mathbf{D}$ makes it almost certain that its eigenvalues have mixed signs so that it is highly likely that the matrix $\tau\mathbf{Q}$ is indefinite. Thus the objective function cannot be guaranteed convex. Such non-convexity means that routines like Ravindran's (1972) do not necessarily converge and that even if an answer is found no guarantee of a global minimum can be given (see for example Ravindran and Lee, 1981). Hansen (1972) has developed an implicit enumeration scheme which does not rely on the convexity of the objective function.

Simple heuristics are clearly applicable to this model, and in view of the size of any realistic application it is likely that this will be the only practical approach. That is, even the most powerful of the above mentioned techniques will not be able to handle problems of realistic size and so a cruder approach is needed. Note that this research takes a somewhat conservative approach to the use of heuristics; it is simple to see how the entire range of STINGY, GREEDY and INTERCHANGE heuristics (Hansen, Peeters and Thisse, 1983) could be adapted to the present case, thereby reducing the number of evaluations needed. Rather than cloud the computational results by introducing two sources of suboptimality (location as well as allocation), we maintain the complete evaluation of locational patterns while investigating the impact of two different assumptions about allocation. An upper bound on the objective function can be found by complete enumeration of the locational configurations with the additional assumption that each node is allocated to its nearest hub. The allocation matrix is thus a by-product of the location patterns (as in the PMP), and so the objective function can be evaluated. (In the notes that follow this heuristic will be called the "nearest hub allocation rule" or HEUR1.)

In one special case it is trivial to show that HEUR1 generates an exact solution. In the case of $\tau=0$ the quadratic term drops out of the QP. Furthermore, it is known that the linear version of the problem is an uncapacitated PMP and this problem has the property that cities are allocated simply once the facility locations are given, using a nearest hub rule. Thus, in the special case of the parameter $\tau=0$, the full quadratic program should simply confirm that the results from the complete HEUR1 evaluation are correct. It is not known a priori how large the value of $\tau$ has to be in order for the nearest hub rule to be invalid. Numerical evidence will be presented for a small case study in the following sections. A tighter bound might be obtained from a slightly more sophisticated heuristic which examines the allocation of every non-hub node to its first or second nearest hub. (In the notes below this heuristic will be called HEUR2.) Since p can be chosen from n nodes in $n!/p!(n-p)!$ ways and for each of these locational patterns there are $(n-p)$ nodes to be assigned to either their first or second nearest facility. There are therefore $(p^{n-p})n!/(n-p)!p!$ to be analyzed. For example with n=10 and p=4, there are 13440 patterns to be analyzed.

9.9 Computational Example - Data and Experimental Design

The interaction data (described above) were arbitrarily divided into the four separate interaction tables, involving 10x10, 15x15, 20x20 and 25x25 flows. (The sub-tables were simply created by taking the top left hand submatrix from the entire 25x25 system.) The analytical design in this phase of the research is to (1) use the HEUR1 procedure to find locations for 2, 3 and 4 hubs for the 10x10, 15x15, 20x20 and 25x25 interaction systems;

(2) use HEUR2 to provide a complete enumeration for the siting of two hubs to service 10 nodes; and (3) use the HEUR2 procedure to find 4 hubs for the 10 node problem.

### 9.9.1   Results using HEUR1

Recall that HEUR1 assumes that each node is assigned to its nearest hub. This assumption makes it possible to enumerate all locational configurations for small problems. Note that by far the most time consuming aspect of the   problem is the evaluation of the quadratic term. Instead, two short cut methods are analyzed the first tests whether a pair of nodes i and j are connected to the same hub; if they are then no quadratic calculation is needed and the procedure can skip on to the next pair. If the places are not connected to the same hub, then the appropriate inter-hub cost can be evaluated. This method requires an intermediate loop to check on the hubs servicing the pair of places but this can easily be carried out in n operations, as follows:  Let $S_{ij} = \sum_k X_{ik} X_{jk}$ . Then the pair i,j is serviced by different hubs, if $S_{ij} = 0$. A second method is based on a set $J$ containing the indices of the open facilities. The objective can then be calculated restricting the summations to the elements of $J$ greatly reducing the number of calculations necessary, at the expense of extra bookkeeping to maintain the index set $J$.

The results shown in Table 9.3 are obtained from routines written in FORTRAN and run on an IBM 3081/D. The data reported in Table 9.3 include the volume of flow in the appropriate sub-matrix (i.e. the sum of the raw values in the interaction matrix) and the objective function value (i.e the quantity in equation (9.7)). The total cost of operating a hub system for a given interaction matrix is obtained by multiplying the volume of flow by the objective function value (recall that the $T$ values are scaled in the computation of the objective in equation (9.7)). For any given interaction system (e.g. 15x15), the objective function value drops as more hubs are opened. Of course, the total cost of operating any given number of hubs increases as these hubs serve larger interaction systems. Note also the non-incremental nature of the solutions: the best known 3-hub system for the 15x15 model opens (4, 5, 11) as hubs; the best known 4-hub system for the same system opens (4, 5, 7, 8). These comments are based on the results of a heuristic method   so that the usual caution must be exercised in interpretation; an optimal method could find better answers. With regard to the two alternative methods of evaluating the objective function, it appears that method 2 is faster when the number of hubs is small relative to the number of interacting nodes. When the number of hubs is large relative to the size of the system (e.g. p=4 for the 10x10 model) the overhead in maintaining the pointers outweighs the savings from the relatively small   reduction in the number of summation terms.

### 9.9.2   Results using HEUR2 for Complete Evaluation of 2 Hubs

Applying HEUR2 to the 10x10 interaction system to find 2 hub locations, necessarily constitutes a complete evaluation, since every possible assignment of the 8 non-hub cities to the two hubs are considered. The best pair of locations occurs at Chicago and Dallas-Fort Worth, for an objective function value of Z=835.8105; (compare the value of 849.2571 in Table 9.3 which was obtained using HEUR1). The cities in the system  do not follow a nearest hub allocation rule - note that Denver is allocated to Chicago which is not its nearest center. It is not computationally feasible to work out complete solutions for any larger system, since the number of location and allocation patterns is
$n!/(n-p)!p!$ $(p)^{n-p}$ , which is over 800,000 for 4 hubs serving 10 nodes.

Table 9.3  **Application of Program to 4 Interaction Systems**

|  | 10x10 | 15x15 | 20x20 | 25x25 |
|---|---|---|---|---|
| Total flow | 999,026 | 2,364,942 | 5,754,594 | 8,540,006 |
| **2 HUBS** | | | | |
| Objective | 849.2571 | 1222.6477 | 1214.3745 | 1365.2141 |
| Locations | Dallas | Chicago | Chicago | Chicago |
|  | Detroit | Kansas City | Pittsburgh | Pittsburgh |
| Evaluations | 45 | 105 | 190 | 300 |
| CPU Time | | | | |
| Method 1 | 0:00.89 | 0:02.79 | 0:09.73 | 0:28.15 |
| Method 2 | 0:00.82 | 0:01.68 | 0:04.05 | 0:09.01 |
| **3 HUBS** | | | | |
| Objective | 791.1733 | 1185.7010 | 1160.7342 | 1284.4979 |
| Locations | Chicago | Chicago | Chicago | Denver |
|  | Dallas | Cincinnati | Kansas | Pittsburgh |
|  | Detroit | Kansas | Pittsburgh | St. Louis |
| Evaluations | 120 | 455 | 1140 | 2300 |
| CPU Time | | | | |
| Method 1 | 0:01.63 | 0:11.95 | 1:04.54 | 4:03.73 |
| Method 2 | 0:01.84 | 0:10.33 | 0:43.27 | 2:13.46 |
| **4 HUBS** | | | | |
| Objective | 739.0123 | 1129.8315 | 1119.0376 | 1220.4011 |
| Locations | Chicago | Chicago | Chicago | Chicago |
|  | Dallas | Cincinnati | Cleveland | Dallas |
|  | Denver | Dallas | Kansas | Denver |
|  | Detroit | Denver | Philadelphia | Pittsburgh |
| Evaluations | 210 | 1365 | 4845 | 12650 |
| CPU Time | | | | |
| Method 1 | 0:02.40 | 0:35.62 | 4:42.92 | >22 mins |
| Method 2 | 0:04.05 | 0:49.27 | 5:07.20 | 20:49.63 |

NOTES: Computations are in minutes:seconds for an IBM 3081/D
CPU Time includes compilation, execution and input-output operations.

Source: O'Kelly (1987b)

### 9.9.3  Results using HEUR2

Applying HEUR2 to the 10x10 interaction system to find 4 hub locations, requires the evaluation of 13440 patterns. At the best known solution (BKS) from HEUR2, locations are opened at Atlanta (1), Chicago (4), Dallas-Fort Worth (7) and Detroit (9) or an objective of Z=736.2594. The cities in the system do not follow a nearest hub allocation rule.
Notice that if the 10 demand nodes are assigned to the nearest hubs using locations 1,4,7 and 9 (i.e. the same hubs as in the BKS), then the objective function value is 746.747. The best answer found from patterns restricted to nearest hub assignment is 739.012 (compare with the results from HEUR1). This example clearly illustrates the lack of separability of the assignment and location phases of the problem. The implied set of interfacility linkages at the best known answer is shown in Figure 9.6A. Hubs are opened at Atlanta, Chicago, Detroit and Dallas-Fort Worth. These hubs are interlinked and routes are then formed by assigning each origin to its allocated hub, proceeding from there to the hub allocated to the required destination, and on ultimately to the destination city.

**Table 9.4  Comparison of HEUR1 and HEUR2 and Impacts of Scale Economies on Solutions**

|     | HEUR1    | HEUR2    | (1,4,7,9) | N <1.02 | Nearest |
|-----|----------|----------|-----------|---------|---------|
| 1.0 | 739.0123 | 736.2594 | 736.2594  | 19      | 6       |
| 0.8 | 661.8285 | 661.4154 | 666.3363  | 19      | 5       |
| 0.6 | 577.8313 | 577.8313 | 596.4131  | 9       | 3       |
| 0.4 | 493.7939 | 493.7939 | 526.4900  | 13      | 3       |
| 0.2 | 395.1305 | 393.1305 | 456.5669  | 3       | 1       |
| 0.0 | 295.3057 | 295.3057 | 386.6438  | 3       | 1       |

HEUR1 the objective function obtained assuming allocation of cities to the nearest facility.

HEUR2 the objective function obtained from HEUR2.

(1,4,7,9) is the solution obtained by holding the facilities at the location and allocation pattern shown in Fig 9.6A.

N < 1.02 is the number of patterns found with an objective less than 2% more than the BKS.

Nearest is the number of patterns found with nearest hub  assignment and with an objective less than 2% more than the BKS.

Source: O'Kelly (1987b)

The patterns discussed so far are generated from an interaction scheme which does not enjoy any scale economies, as the parameter τ=1.0 . Now the model is applied to the same data for five other parameter values τ=0.8, 0.6, 0.4, 0.2 and 0.0. A comparison of objective functions from HEUR1 and HEUR2 is shown in Table 9.4. Five pieces of information are provided for each $\tau$ value: the nearest hub allocation result (HEUR1); the best known solution (from HEUR2); the objective evaluated at the location-allocation pattern shown in Figure 9.6A; the number of solutions found within 2% of the BKS; and finally the number of these "good" solutions which involve nearest hub allocation. The answers are intuitively reasonable: as the discount increases (i.e. $\tau$ decreases) the difference between the nearest hub assignment answer and the best known answer gets smaller. If the locations are fixed at those chosen using $\tau=1.0$, progressively larger percentage penalties will be incurred as $\tau$ decreases. The number of solutions found to be within 2% of the BKS decreases as $\tau$ decreases.

The spatial results are shown in Figure 9.6B for $\tau=0.8$, Figure 9.6C for $\tau=0.6$ and $\tau=0.4$, and Figure 9.6D for $\tau=0.2$ and $\tau=0.0$. Recall that $\tau=0.0$ corresponds to the linear version of the problem. As $\tau$ decreases the interfacility linkages become less expensive per unit distance than segments of routes which include one non-hub city (i.e. spokes). When $\tau=0.8$ facilities are opened in Chicago, Dallas-Fort Worth, Denver and Detroit. The topology of the system has changed with 2 cities assigned to Chicago, 1 to Dallas, 0 to Denver and 3 to Detroit. Therefore, from the formulae developed for the numbers of routes (O'Kelly, 1986b), it is found that there are still 24 direct connections, 44 one stop and 22 two stop routes. The objective function value at the BKS is Z=661.42. It turns out that a solution with a very similar objective function value is found if the fourth hub is placed in Cleveland instead of Detroit. Further analyses of the numbers and types of routes are reported in O'Kelly (1986b).

## 9.10  Conclusions

This chapter has introduced the problem of locating hub facilities so as to serve a set of interacting cities. The importance of spatial interaction in determining the optimal location for hubs in networks was demonstrated. It is known that this problem has a large range of applications in the areas of air transportation networks, communications systems, and in any distribution task which requires efficient connections between nodes. The major rationale for a hub system is that it reduces the number of routes which must be operated. This research has shown that the planar version of the single-hub minisum location problem is identical to finding a Weberian least cost site. The single-hub minimax problem is similar to the round trip location model of Chan and Hearn (1977). Networks organised around two hubs are more difficult to analyze because routing (assignment) of demand nodes to hubs must be determined. As the number of assignments increases exponentially in the number of nodes, complete enumeration is out of the question.

Figure 9.6  (A,B,C,D) Analysis of Scale Effects - Best Solutions

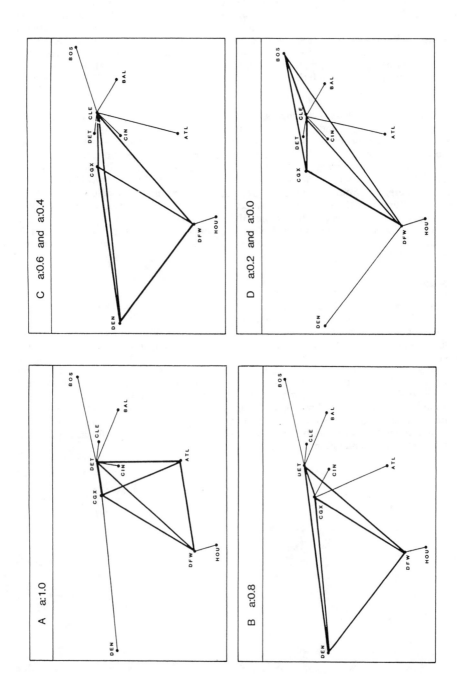

A simplifying assumption is made; only the non-overlapping partitions of the nodes between the two facilities are examined. While this assumption actually generates the optimal solution for a two center location-allocation model (Ostresh, 1975), no such guarantee can be given in the present case. Recent research by Aykin (1987) has followed up on this problem and shows that both the number of cities assigned to each hub, as well as the relative magnitudes of the intercity flows, are important to the validity of the assumption.

The first two parts of this chapter discussed planar location models, but it is clear that in many real-world applications the choice of hub location is limited to a fixed set of sites. Therefore, part 3 of the chapter introduced a quadratic programming model of interacting hub facility location. This problem is shown to be more difficult than the related p-median problem because of the lack of a simple rule for solving the allocation phase of the program. Thus, a full enumeration would involve the solution of a large number of quadratic assignment problems. In view of the difficulty of the mathematical program, we have attacked the problem with some heuristic procedures. It appears that the nature of the quadratic hub location problem is sufficiently difficult that crude heuristics (like HEUR1 which assumes nearest center assignment) will be the only practical approach to solution. The results in this chapter have shown that for moderate savings in inter-hub linkage costs (i.e. $\tau \leq 0.5$) HEUR1 (nearest hub rule) gives an answer similar to the more time-consuming HEUR2. For problems with $\tau > 0.5$ it is advisable to check at least the second possible assignment and for small problems this can be carried out by complete enumeration.

# LITERATURE CITED

Achen, C.H. (1982) *Interpreting and Using Regression.* Quantitative Applications in the Social Sciences, 29, Sage: Beverly Hills.

Adler, T. and M. Ben-Akiva (1979) A theoretical and empirical model of trip chaining behavior, *Transportation Research,* 13B:243-257.

Alexis, M., G. Haines and L. Simon (1968) Consumer information processing: the case of women's clothing, in *Marketing and the New Science of Planning,* R.L. King (ed.), American Marketing Association: Chicago.

Allard, L. and M.J. Hodgson (1987) Interactive graphics for mapping location-allocation solutions, *American Cartographer,* 14(1):49-60.

Alonso, W. (1973) *National Interregional Demographic Accounts: A Prototype.* Monograph 17, Institute of Urban and Regional Development, University of California, Berkeley.

Alonso, W. (1978) A theory of movement, in *Human Settlement Systems,* N.M. Hansen (ed.), Cambridge: Ballinger, pp. 197-211.

Ambrose, D.M. (1979) Retail grocery pricing: inner city, suburban and rural comparisons, *Journal of Business,* 52(1):95-102.

Ambrose, P.J. (1968) An analysis of intra-urban shopping patterns, *Town Planning Review,* 38:327-334.

Andrews, P.W.S. and E. Brunner (1975) *Studies in Pricing.* Macmillan: London.

Angel, S. and G. Hyman (1976) *Urban Fields.* Pion: London.

Artle, R. (1959) *Studies in the Structure of the Stockholm Economy.* Business Research Institute: Stockholm.

Aykin, T. (1987) On the location of hub facilities, *Transportation Science,* 22(2):155-157.

Bacon, R.W. (1971) An approach to the theory of consumer shopping behaviour, *Urban Studies,* 8:55-64.

Bacon, R.W. (1984) *Consumer Spatial Behaviour: A Model of Purchasing Decisions over Space and Time.* Oxford University Press: London.

Batsell, R.R. (1981) A multiattribute extension of the Luce model which simultaneously scales utility and substitutability, Working Paper, J.H. Jones Graduate School of Administration, Rice University, Texas.

Batsell, R.R. and J.C. Polking (1985) A generalized model of market share, *Marketing Science,* 4:62-73.

Batty, M. (1976a) *Urban Modelling: Algorithms, Calibrations, Predictions.* Cambridge University Press: Cambridge, U.K..

Batty, M. (1976b) Entropy in spatial aggregation, *Geographical Analysis,* 8:1-21.

Batty, M. (1978) Reilly's challenge: new laws of retail gravitation which define systems of central places, *Environment and Planning A,* 10:185-219.

Batty, M. and S. Mackie (1972) The calibration of gravity, entropy, and related models of spatial interaction, *Environment and Planning A,* 4:205-233.

Batty, M. and L. March (1976) The method of residues in urban modelling, in *Colloquium on Socio-Economic Systems Design in Urban and Regional Planning,* University of Waterloo, pp. 28-29.

Baumol, W. and I. Ide (1956) Variety in retailing, *Management Science,* 3:93-101.

Baxter, M.J. (1984) A note on the estimation of a nonlinear migration model, *Geographical Analysis,* 16:282-286.

Baxter, M.J. and G.O. Ewing (1981) Models of recreational trip distribution, *Regional Studies,* 15:327-344.

Beaumont, J.R. (1980) Spatial interaction models and the location-allocation problem, *Journal of Regional Science,* 20(1):37-50.

Beaumont, J.R. (1981) Location-allocation models in a plane: a review of some models, *Socio-Economic Planning Sciences,* 15(5):217-229.

Beavon, K. and A. Hay (1977) Consumer choice of shopping centre - a hypogeometric model, *Environment and Planning A,* 9:1375-1393.

Beckenbach, E.F. and R. Bellman (1961) *An Introduction to Inequalities.* Random House: New York.

Beckmann, M.J. (1983) On optimal spacing under exponential distance effect, in *Locational Analysis of Public Facilities,* J.-F. Thisse and H. Zoller (eds.), North-Holland: Amsterdam, pp. 117-125.

Bell Telephone Laboratories (1977) *Engineering and Operations in the Bell System.* Library of Congress Catalog Card No. 77-84418

Bellman, R. (1970) *Introduction to Matrix Analysis.* 2nd edition. McGraw-Hill: New York.

Ben-Akiva, M.F. and S.R. Lerman (1985) *Discrete Choice Analysis.* MIT Press: Cambridge, MA.

Bettman, J.R. (1979) *An Information Processing Theory of Consumer Choice.* Addison-Wesley: Reading, MA.

Bishop, Y.M.M., S.E. Feinburg and P.W. Holland (1975) *Discrete Multivariate Analysis: Theory and Practice.* MIT Press: Cambridge, MA.

Blanco, C. (1963) The determinants of interstate population movements, *Journal of Regional Science,* 5:77-84.

Boal, F.W. (1969) Territoriality on the Shankill-Falls divide, Belfast, *Irish Geography,* 6(1):30-50.

Borgers, A. and H.J.P. Timmermans (1986) A model of pedestrian route choice and demand for retail facilities within inner-city shopping areas, *Geographical Analysis,* 18:115-128.

Borgers, A. and H.J.P. Timmermans (1987) Choice model specification, substitution and spatial structure effects: a simulation experiment, *Regional Science and Urban Economics,* at press.

Brady, S.D. and R.E. Rosenthal (1980) Interactive computer graphical solutions of constrained minimax location problems, *AIIE Transactions,* 12(3):241-248.

Broadbent, T.A. (1970) Notes on the design of operational models, *Environment and Planning,* 2:469-476.

Brotchie, J.F. (1969) A general planning model, *Management Science,* 16:265.

Brown, A.J. and E. M. Burrows (1977) *Regional Economic Problems.* George Allen and Unwin: London.

Burnett, P. (1973) The dimensions of alternatives in spatial choice processes, *Geographical Analysis,* 5:181-204.

Burnett, P. (1974) A Bernoulli model of destination choice, *Transportation Research Record,* 527:33-44.

Burnett, P. (1977) Tests on a linear learning model of destination choice: applications to shopping travel by heterogeneous population, *Geografiska Annaler B,* 59:95-108.

Burnett, P. (1978) Markovian models of movement within urban spatial structures, *Geographical Analysis,* 10:142-153.

Burnett, P. and S. Hanson (1979) A rationale for an alternative mathematical paradigm for movement as complex human behavior, *Transportation Research Record,* 723:11-24.

Burns, L. (1979) *Transportation, Spatial and Temporal Components of Accessibility.* Lexington Books: MA.

Bussiere, R. and F. Snickars (1970) Derivation of the negative exponential model by an entropy maximising method, *Environment and Planning,* 2:295-301.

Cadwallader, M.T. (1975) A behavioral model of consumer spatial decision making, *Economic Geography,* 51:339-349.

Cadwallader, M.T. (1979) Neighbourhood evaluation in residential mobility, *Environment and Planning A,* 11:393-402.

Cadwallader, M.T. (1981) Towards a cognitive gravity model: the case of consumer spatial behavior, *Regional Studies*, 15:275-284.

Carlstein, T., D. Parkes and N. Thrift (1978) *Human Activity and Time Geography*. Wiley: New York.

Cebula, R.J. and R.K. Vedder (1973) A note on migration, economic opportunity, and the quality of life, *Journal of Regional Science*, 13:205-212.

Cesario, F. (1975a) A primer on entropy modeling, *Journal of the American Institute of Planners*, 41:40-48.

Cesario, F. (1975b) Linear and non-linear regression models of spatial interaction, *Economic Geography*, 51:69-77.

Champion, A.G., Green, A.E., Owen, D.W., Ellin, D.J. and M.G. Coombes (1987) *Changing Places: Britain's Demographic, Economic and Social Complexion*. Edward Arnold: London.

Chan, A.W. and D.W. Hearn (1977) A rectilinear distance round-trip location problem, *Transportation Science*, 11:107-123.

Chan, Y. and R.J. Ponder (1979) The small package air freight industry in the United States: a review of the Federal Express experience, *Transportation Research*, 13A:221-229.

Cheney, W. and D. Kincaid (1980) *Numerical Mathematics and Computing*. Brooks/Cole: Monterey.

Choukroun, J.-M. (1975) A general framework for the development of gravity-type trip distribution models, *Regional Science and Urban Economics*, 5:177-202.

Clark, G.L. and K.P. Ballard (1980) Modeling out-migration from depressed regions: the significance of origin and destination characteristics, *Environment and Planning A*, 12:799-812.

Clark, W.A.V. (1968) Consumer travel patterns and the concept of range, *Annals of the Association of American Geographers*, 58:386-396.

Clark, W.A.V. and G. Rushton (1970) Models of intra-urban consumer behaviour and their implications for central place theory, *Economic Geography*, 46:486-496.

Clark, W.A.V. and M.T. Cadwallader (1973) Residential preferences: an alternative view of intraurban space, *Environment and Planning A*, 5:693-703.

Coombes, M.G., J.S. Dixon, J.B. Goddard, S. Openshaw and P.J. Taylor (1982) Functional regions for the population centres of Britain, in *Geography and the Urban Environment 5*, D.T. Herbert and R.J. Johnston (eds.), Wiley: Chichester.

Cooper, L. (1963) Location-allocation problems, *Operations Research*, 11:331-343.

Cooper, R.A. and A.J. Weekes (1983) *Data, Models and Statistical Analysis*. Barnes and Noble: Totowa, NJ.

Cordey-Hayes, M. and A.G. Wilson (1971) Spatial interaction, *Socio-Economic Planning Sciences*, 5:73-95.

Cox, D.R. (1970) *The Analysis of Binary Data*. Methuen: London.

Crafton (1979) Convenience store pricing and the value of time: a note on the Becker-Devany full price model, *Southern Economic Journal*, 45(4):1254-1260.

Cyert and March (1963) *A Behavioral Theory of the Firm*. Prentice Hall: Englewood Cliffs NJ.

Dacey, M.P. and A. Norcliffe (1977) A flexible doubly-constrained trip distribution model, *Transportation Research*, 11B:203-204.

Damm, D. (1979) *Towards a model of activity scheduling behavior*. Unpublished Ph.D. Thesis, Department of Civil Engineering and Urban Studies, M.I.T.: Cambridge MA.

Damm, D. (1980) Interdependencies in activity behavior, *Transportation Research Record*, 750:33-40.

Damm, D. (1982) Parameters of activity behavior for use in travel analysis, *Transportation Research*, 16A:135-148.

Damm, D. (1983) Theory and empirical results: a comparison of recent activity-based research, in *Recent Advances in Travel Demand Analysis*, S. Carpenter and P. Jones (eds.), Gower.

Damm, D. and S. Lerman (1981) Theory of activity scheduling behavior, *Environment and Planning A*, 13:703-718.

Daniels, P.W. and A.M. Warnes (1980) *Movement in Cities: Spatial Perspectives on Urban Transportation and Travel*. Methuen: New York.

Daskin, M. (1987) Location, dispatching, and routing models for emergency services with stochastic travel times, in *Spatial Analysis and Location-Allocation Models*, A. Ghosh and G. Rushton (eds.), van Nostrand Reinhold: New York, pp. 224-265.

Davies, R.B. and R. Crouchley (1985) Control for omitted variables in the analysis of panel and other longitudinal data, *Geographical Analysis*, 17:1-15.

Davies, R.B. and C.M. Guy (1987) The statistical modeling of flow data when the poisson assumption is violated, *Geographical Analysis*, 19:300-314.

Davies, R.B. and A.R. Pickles (1987) A joint trip timing store-type choice model for grocery shopping, including inventory effects and nonparametric control for omitted variables, *Transportation Research*, 21A:345-361.

Davies, R.L. (1969) Effects of consumer income differences on shopping movement behaviour, *Tijdschrift voor Economische en Social Geografie*, 60:111-121.

Davies, R.L. (1976) *Marketing Geography*. Methuen:London.

Daws, L.B. and A.J. Bruce (1971) *Shopping in Watford.* Building Research Establishment: Garston UK.

Deacon et al. (1972) Models of outdoor recreational travel, *Highway Research Record,* 472:28-44.

Devine, D.G. and B.W. Marion (1979) The influence of consumer price information on retail pricing and consumer behavior, *American Journal of Agricultural Economics,* 61(2):228-237.

Dorigo, G. and W. Tobler (1983) Push-pull migration laws, *Annals of the Association of American Geographers,* 73(1):1-17.

Downs, R. M. (1970) The cognitive structure of an urban shopping center, *Environment and Behavior,* 2:13-39.

Dowson, D.C. and A. Wragg (1973) Maximum-entropy distributions having prescribed first and second moments, *IEEE Transactions on Information Theory,* IT-19:689-693.

Drezner, Z. (1982) Fast algorithms for the round trip location problem, *IIE Transactions,* 14(4):243-248.

Dunlevy, J.A. and H.A. Gemery (1977) The role of migrant stock and lagged migration in the settlement patterns of nineteenth century immigrants, *Review of Economics and Statistics,* 59:137-144.

Eaton, B.C. and R.G. Lipsey (1976) The introduction of space into the neoclassical model of value theory, Discussion Paper 239, Inst. for Economic Research, Queen's University, Kingston, Ontario.

Eaton, B.C. and R.G. Lipsey (1982) An economic theory of central places, *Economic Journal,* 92:56-72.

Eilon, S., C.D.T. Watson-Gandy and N. Christofides (1971) *Distribution Management: Mathematical Modelling and Practical Analysis.* Griffin: London.

Elzinga, J. and D.W. Hearn (1972) Geometrical solutions for some minimax location problems, *Transportation Science,* 6:379-394.

Erlenkotter, D. and G. Leonardi (1985) Facility location with spatially-interactive behavior, *Sistemi Urbani,* 1:29-41.

Evans, A.W. (1970) Some properties of trip distribution models, *Transportation Research,* 4:19-36.

Evans, R.A. (1969) The principle of minimum information, *IEEE, Transactions on Reliability,* R-18:87-90.

Evans, S. (1973) A relationship between the gravity model for trip distribution and the transportation problem in linear programming, *Transportation Research,* 7:39-61.

Evans, S. (1976) Derivation and analysis of some models for combining trip distribution and assignment, *Transportation Research,* 10:37-57.

Ewing, G.O. (1976) Environmental and spatial preferences of interstate migrants in the United States, in *Spatial Choice and Spatial Behavior: Geographic Essays on the Analysis of Preferences and Perceptions*, R.G. Golledge and G. Rushton (eds.), Ohio State University Press: Columbus, pp. 249-270.

Fingleton, B. (1975) A factorial approach to the nearest centre hypothesis, *Transactions of the Institute of British Geographers*, 65:131-139.

Fisher, R.A. and L.H.C. Tippett (1928) Limiting forms of the frequency distribution of the largest or smallest member of a sample, *Proceedings of the Cambridge Philosophical Society*, 24:180-190.

Flowerdew, R. and M. Aitkin (1982) A method of fitting the gravity model based on the Poisson distribution, *Journal of Regional Science*, 22:191-202.

Flowerdew, R. and J. Salt (1979) Migration between labour market areas in Great Britain, 1970-1971, *Regional Studies*, 13:211-231.

Fotheringham, A.S. (1981) Spatial structure and distance-decay parameters, *Annals of the Association of American Geographers*, 71:425-436.

Fotheringham, A.S. (1983a) A new set of spatial interaction models: the theory of competing destinations, *Environment and Planning A*, 15:15-36.

Fotheringham, A.S. (1983b) Some theoretical aspects of destination choice and their relevance to production-constrained gravity models, *Environment and Planning A*, 15:1121-1132.

Fotheringham, A.S. (1984) Spatial flows and spatial patterns, *Environment and Planning A*, 16:529-543.

Fotheringham, A.S. (1985) Spatial competition and agglomeration in urban modelling, *Environment and Planning A*, 17:213-230.

Fotheringham, A.S. (1986) Modelling hierarchical destination choice, *Environment and Planning A*, 18:401-418.

Fotheringham, A.S. (1987) Hierarchical destination choice: discussion with evidence from migration in The Netherlands, Working Paper 69, Netherlands Interuniversity Demographic Institute: The Hague.

Fotheringham, A.S. (1988a) Market share analysis techniques: a review and illustration of current U.S. practice, in *Store Choice, Store Location and Market Analysis*, N. Wrigley (ed.), Routledge: London, pp. 120-159.

Fotheringham, A.S. (1988b) Consumer store choice and choice set definition, *Marketing Science*, 7(3) forthcoming.

Fotheringham, A.S. (1988c) Spatial information processing and spatial choice modelling, manuscript.

Fotheringham, A.S. and T. Dignan (1984) Further contributions to a general theory of movement, *Annals of the Association of American Geographers*, 74:620-633.

Fotheringham, A.S. and D.C. Knudsen (1984) Critical parameter values in retail shopping models, *Modeling and Simulation* 15, part 1: 75-80, Instrument Society of America.

Fotheringham, A.S. and D.C. Knudsen (1986) Modelling discontinuous change in the spatial pattern of retail outlets: a methodology in *Transformations Through Space and Time: An Analysis of Nonlinear Structures, Bifurcation Points and Autoregressive Dependencies,* D.A. Griffith and R.P. Haining (eds.), Martinus Nijhoff: Boston, pp. 273-292.

Fotheringham, A.S. and D.C. Knudsen (1987) *Goodness-of-Fit Statistics.* Concepts and Techniques in Modern Geography, 46, Geo Books: Norwich.

Fotheringham, A.S. and R. Flowerdew (1988) Comments on the spatial choices of British migrants, manuscript.

Fotheringham, A.S. and P.A. Williams (1983) Further discussion on the Poisson interaction model, *Geographical Analysis,* 15:343-347.

Francis, R. and J.A. White (1974) *Facility Layout and Location.* Prentice-Hall: Englewood Cliffs, N.J.

Gale, S. (1972) Inexactness, fuzzy sets, and the foundations of behavioral geography, *Geographical Analysis,* 4:337-349.

Gallo, G., P.L. Hammer and B. Simeone (1980) Quadratic knapsack problems, in *Combinatorial Optimization, Mathematical Programming Study,* 12, M.W. Padberg (ed.), North Holland: Amsterdam.

Garrison, W.L. et al. (1959) *Studies of Highway Development and Geographic Change.* University of Washington Press: Seattle.

Gautschi, D.A. (1981) Specification of patronage models for retail center choice, *Journal of Marketing Research,* 18:162-181.

General Household Survey (1982), Her Majesty's Stationary Office: London.

Georgescu-Roegen, N. (1971) *The Entropy Law and the Economic Process.* Harvard University Press: Cambridge.

Ghosh, A. and S. McLafferty (1984) A model of consumer propensity for multipurpose shopping, *Geographical Analysis,* 16:244-249.

Gilbert, G., G.L. Peterson and J.L. Schofer (1972) Markov renewal model of linked travel behavior, *Transport Engineering Journal,* 98:691-704.

Giles, D.A. and P. Hampton (1981) Interval estimation in the calibration of certain trip distribution models, *Transportation Research B,* 15B:203-219.

Goldman and Johansson (1978) Determinants of search for lower prices: an empirical assessment of the economics of information, *Journal of Consumer Research,* 5:176-186.

Golledge, R. and A. Spector (1978) Comprehending the urban environment: theory and practice, *Geographical Analysis,* 10(4):403-426.

Goodchild, M.F. (1984) ILACS: a location-allocation model for retail site selection, *Journal of Retailing*, 60(1):84-100.

Goodchild, M.F. and P.J. Booth (1976) Modelling human spatial behaviour in urban recreation facility site location, DP-007, Dept. of Economics, Univ. of Western Ontario.

Gould, P. (1972) Pedagogic review, *Annals of the Association of American Geographers*, 62:689-700.

Gould, P. and R. White (1974) *Mental Maps*. Allen and Unwin: Boston, MA.

Gray, R.H. and A.K. Sen (1983) Estimating gravity model parameters: A simplified approach based on the odds ratio, *Transportation Research B*, 17:117-131.

Greenwood, M.J. (1969) The determinants of labor migration in Egypt, *Journal of Regional Science*, 9:283-290.

Greenwood, M.J. (1971) A regression analysis of migration to urban areas of a less-developed country: the case of India, *Journal of Regional Science*, 11:253-264.

Greenwood, M.J. and D. Sweetland (1972) The determinants of migration between Standard Metropolitan Statistical Areas, *Demography*, 9:665-681.

Griffith, D.A. (1982) A generalized Huff model, *Geographical Analysis*, 14(2):135-144.

Gujarati, D. (1978) *Basic Econometrics*. McGraw-Hill: New York.

Guy, C.M. and N. Wrigley (1987) Walking trips to shops in British cities, *Town Planning Review*, 58(1):63-79.

Haining, R. (1985) The spatial structure of competition and equilibrium price dispersion, *Geographical Analysis*, 17(3):231-242.

Hallefjord, A. and K. Jornsten (1985) A note on relaxed gravity models, *Environment and Planning A*, 17:597-603.

Handler, G.Y. and P. Mirchandani (1979) *Location on Networks: Theory and Algorithms*. MIT Press: Cambridge MA.

Hansen, P. (1972) Quadratic zero-one programming by implicit enumeration, in *Numerical Methods for Non-linear Optimization*, F.A. Lootsma (ed.), Academic Press: New York, pp. 265-278.

Hansen, P. (1979) Methods of nonlinear 0-1 programming, in *Discrete Optimization II*, P.L. Hammer, E.L. Johnson and B.H. Korte (eds.), North Holland: Amsterdam, pp. 53-70.

Hansen, P., D. Peeters, D. Richard and J.-F. Thisse (1985) The minisum and minimax location problems revisited, *Operations Research*, 33(6):1251-1265.

Hansen, P., D. Peeters and J.-F. Thisse (1983) Public facility location models: a selective survey, in *Locational Analysis of Public Facilities*, J.-F. Thisse and H. Zoller (eds.), North Holland: Amsterdam, pp. 223-262.

Hansen, R.A. and T. Deutscher (1977) An empirical investigation of attribute importance in retail store selection, *Journal of Retailing*, 53:59-72.

Hansen, W.G. (1959) How accessibility shapes land use, *Journal of the American Institute of Planners*, 25:73-76.

Hanson, S. (1977) Measuring the cognitive levels of urban residents, *Geografiska Annaler*, 59B:67-87.

Hanson, S. (1979) Urban-travel linkages: a review, in *Behavioral Travel Modelling*, D. Hensher and P. Stopher (eds.), Croom Helm: London, pp. 81-100.

Hanson, S. (1980a) Spatial diversification and multipurpose travel: implications for choice theory, *Geographical Analysis*, 12:245-257.

Hanson, S. (1980b) The importance of the multipurpose journey to work in urban travel behavior, *Transportation*, 9:229-248.

Hanson, S. (1986) *The Geography of Urban Transportation*. Guilford Press: New York.

Hanushek, E.A. and J.E. Jackson (1977) *Statistical Methods for Social Scientists*. Academic Press: New York.

Harris, B. and A.G. Wilson (1978) Equilibrium values and dynamics of attractiveness terms in a production-constrained spatial-interaction model, *Environment and Planning A*, 10:371-388.

Harris, C.D. (1954) The market as a factor in the localization of industry in the United States, *Annals of the Association of American Geographers*, 44:315-348.

Hathaway, P.J. (1975) Trip distribution and disaggregation, *Environment and Planning A*, 7:71-97.

Haworth, J.M. and P.J. Vincent (1979) The stochastic disturbance specification and its implications for log-linear regression, *Environment and Planning A*, 11:781-790.

Hay, A.M. and R.J. Johnston (1979) Search and the choice of shopping centre: two models of variability in destination selection, *Environment and Planning A*, 11:791-804.

Hay, A.M. and R.J. Johnston (1980) Spatial variations in grocery prices: further attempts at modelling, *Urban Geography*, 1:189-201.

Haynes, K.E. and A.S. Fotheringham (1984) *Gravity and Spatial Interaction Models*. Sage: Beverly Hills CA.

Heien, P.M. (1968) A note on log-linear regression, *Journal of the American Statistical Association*, 63:1034-1038.

Held, M., P. Wolfe and H.P. Crowder (1974) Validation of subgradient optimization, *Mathematical Programming*, 6:62-88.

Hemmens, G.C. (1970) Analysis and simulation of urban activity patterns, *Socio-Economic Planning Sciences*, 4:53-66.

Hensher, D.A. (1976) The structure of journeys and the nature of travel patterns, *Environment and Planning A*, 8:655-672.

Hensher, D.A. and L.W. Johnson (1981) *Applied Discrete Choice Modelling*. Croom Helm: New York.

Hervitz, H. (1983) *An analysis of the determinants of interregional migration, within the framework of a multinomial logit model*. Unpublished Ph.D. thesis, Department of Economics, Indiana University, Bloomington, Indiana.

Hirsh, M., J.N. Prashker and M. Ben-Akiva (1984) Theoretical model of weekly activity pattern. Paper presented at the 63rd Annual Meeting of the Transportation Research Board, Washington D.C., January 1984. (Preprinted as Pub No 84-049 Transportation Research Institute, Technion: Israel Institute of Technology.)

Hobson, A. (1969) A new theorem of information theory, *Journal of Statistical Physics*, 1:383-391.

Hobson, A. and B.-K. Cheng (1973) A comparison of the Shannon and Kullback information measures, *Journal of Statistical Physics*, 7:301-310.

Hodgart, R.L. (1978) Optimizing access to public services: a review of problems, models, and methods of locating central facilities, *Progress in Human Geography*, 2:17-48.

Hodgson, M.J. (1978) Towards more realistic allocation in location-allocation models: an interaction approach, *Environment and Planning A*, 10:1273-1285.

Hodgson, M.J. (1981) A location-allocation model maximizing consumers welfare, *Regional Studies*, 15(6):493-506.

Hodgson, M.J. (1984) Alternative approaches to hierarchical location-allocation systems, *Geographical Analysis*, 16:275-281.

Hodgson, M.J. (1986) An hierarchical location-allocation model with allocations based on facility size, *Annals of Operational Research*, 6:273-289.

Hodgson, M.J. (1987) An hierarchical location-allocation model for primary health care delivery in a developing area. Paper presented at ISOLDE IV, Namur Belgium.

Hogan, K. and C. ReVelle (1983) Backup coverage concepts in the location of emergency service, *Modeling and Simulation*, 14:1423-1428.

Horowitz, J. (1978) A disaggregate demand model for nonwork travel, *Transportation Research Record*, 592:1-5.

Horowitz, J. (1980) A utility maximizing model of the demand for multi-destination non-work travel, *Transportation Research B*, 14:369-386.

Horowitz, J. (1981) Identification and diagnosis of specification errors in the multinomial logit model, *Transportation Research B,* 15B:345-360.

Horton, F.E. (1968) Location factors as determinants of consumer attraction to retail firms, *Annals of the Association of American Geographers,* 58:787-801.

Horton, F.E. and D.R. Reynolds (1971) Effects of urban spatial structure on spatial behaviour, *Economic Geography,* 47:36-46.

Horton, F.E. and P.W. Shuldiner (1967) The analysis of land-use linkages, *Highway Research Record,* 165:96-107.

Horton, F.E. and W.E. Wagner (1969) A markovian analysis of urban travel behavior: pattern response by socio-economic occupational group, *Highway Research Record,* 283:19-29.

Hotelling, H. (1929) Stability in competition, *Economic Journal,* 39:41-57.

Howard, R. (1971) *Dynamic Probabilistic Systems.* Wiley: New York.

Howe, A. and K. O'Connor (1982) Travel to work and labor force participation of men and women in an Australian metropolitan area, *Professional Geographer,* 34:50-64.

Hua, C. (1980) An exploration of the nature and rationale of a systemic model, *Environment and Planning A,* 12:713-726.

Hubbard, R. (1978) A review of selected factors conditioning consumer travel behavior, *Journal of Consumer Research,* 5:1-21.

Huber, J., J.W. Payne and C. Pluto (1982) Adding a symmetrically dominated alternative: violations of regularity and the similarity hypothesis, *Journal of Consumer Research,* 9:90-98.

Hudson, R. (1974) Images of the retailing environment: an example of the use of the repertory grid methodology, *Environment and Behaviour,* 6:470-494.

Huff, D.L. (1963) A probabilistic analysis of shopping center trade areas, *Land Economics,* 39:81-90.

Huff, J.O. (1986) Geographic regularities in residential search behaviour, *Annals of the Association of American Geographers,* 76:208-228.

Isbell, E.C. (1944) Internal migration in Sweden and intervening opportunities, *American Sociological Review,* 9:627-639.

Ishikawa, Y. (1987) An empirical study of the competing destinations model utilising Japanese interaction data, *Environment and Planning A,* 19:1359--1373.

Jacobsen, S.K. (1987) On heuristics for some entropy maximizing location models, Research Report 12/87, IMSOR, Institute of Mathematical Statistics and Operations Research, Technical Institute of Denmark.

Jaynes, E.T. (1957) Information theory and statistical mechanics, *Physical Review,*

106(4):620-630.

Jaynes, E.T. (1968) Prior probabilities, *IEEE, Transactions on Systems Science and Cybernetics*, SSC-4:227-241.

Jefferson, T.R. and C.H. Scott (1979) The analysis of entropy models with equality and inequality constraints, *Transportation Research*, 13B:123-132.

Johnson, L. and D.A. Hensher (1979) A random coefficient model of the determinants of frequency of shopping trips, *Australian Economic Papers*, 18:322-336.

Johnston, R.J. (1973) Spatial patterns in suburban evaluations, *Environment and Planning A*, 5:385-395.

Johnston, R.L. (1982) *Numerical Methods*. Wiley: New York.

Jornsten, K.O. (1980) A maximum entropy combined distribution and assignment model solved by Benders decomposition, *Transportation Science*, 12(3):262-276.

Joseph, A.E. and D.R. Phillips (1984) *Accessibility and Utilization: Geographical Perspectives on Health Care Delivery*. Harper and Row: New York.

Kahn, B., W.L. Moore and R. Glazer (1985) An experiment in constrained choice, Working Paper, Graduate School of Management, UCLA.

Karlqvist, A. and B. Marksjo (1971) Statistical urban models, *Environment and Planning*, 3:83-98.

Kau, J.B. and C.F. Sirmans (1979) A recursive model of the spatial allocation of migrants, *Journal of Regional Science*, 19:47-56.

Kennington, J.L. and R.V. Helgason (1980) *Algorithms for Network Programming*. Wiley: New York.

Kitamura, R. (1983) A sequential, history-dependent approach to trip-chaining behavior, *Transportation Research Record*, 944:13-22.

Kitamura, R. (1984a) Incorporating trip chaining to analysis of destination choice, *Transportation Research*, 18B:67-18.

Kitamura, R. (1984b) A model of daily time allocation to discretionary out-of-home activities and trips, *Transportation Research*, 18B:255-266.

Kitamura, R. and M. Kermanshah (1985) Sequential model of interdependent activity and destination choices, *Transportation Research Record*, 987:81-89.

Kitamura, R. and T. Lam (1981) A time dependent markov renewal model of trip chaining, in Proceedings of the Eighth International Symposium on Transportation and Traffic Theory, University of Toronto Canada, June 1981.

Knudsen, D.C. and A.S. Fotheringham (1986) Matrix comparison, goodness-of-fit and spatial interaction modelling, *International Regional Science Review*, 10:401-418.

Kondo, K. (1974) Estimation of person trip patterns and modal split, in

*Transportation and Traffic Theory*, D.J. Buckley (ed.), New South Wales University, pp. 715-742.

Koppelman, F.S. and J.R. Hauser (1978) Destination choice behaviour for non-grocery shopping trips, *Transportation Research Record*, 673:157-165.

Laber, G. and R.X. Chase (1971) Interprovincial migration in Canada as a human capital decision, *Journal of Political Economy*, 79:795-804.

Lakshmanan, T.R. and W.A. Hansen (1965) A retail market potential model, *Journal of the American Institute of Planners*, 31:134-143.

Landau, U., J.N. Prashker and B. Alpern (1982) Evaluation of activity constrained choice sets to shopping destination choice modelling, *Transportation Research*, 16A:199-207.

Landau, U., J.N. Prashker and M. Hirsh (1981) The effect of temporal constraints on household travel behavior, *Environment and Planning A*, 13:435-448.

Ledent, J. (1982) Rural-Urban migration, urbanization, and economic development, *Economic Development and Cultural Change*, 30:507-538.

Ledent, J. (1985) The doubly constrained model of spatial interaction: a more general formulation, *Environment and Planning A*, 17:253-262.

Lehtinen, U. (1974) A brand choice model: theoretical framework and empirical results, *European Research*, 2:51-68.

Leonardi, G. (1980) A unifying framework for public facility location problems, WP-80-79, IIASA, Laxenburg, Austria.

Leonardi, G. (1983) The use of random-utility theory in building location-allocation models, in *Locational Analysis of Public Facilities*, J.-F. Thisse and H. Zoller (eds.), North-Holland: Amsterdam, pp. 357-383.

Levine, R. and M. Tribus (1979) *The Maximum Entropy Formalism*. MIT Press: Cambridge MA.

Lewis, P.E. (1975) An empirical test of alternative theories of trade, *Annals of Regional Science*, 9:102-111.

Lewis-Beck, M.S. (1980) *Applied Regression: An Introduction*. Quantitative Applications in the Social Sciences, 22, Sage: Beverly Hills.

Lincoln, D.J. and A.C. Samli (1979) Definitions, dimensions and measurement of store image: a literature summary and synthesis, in *Proceedings of the Southern Marketing Association*, R. Franz, R. Hopkins and A. Toma (eds.), pp. 430-433.

Lindsay, P.H. and D.A. Norman (1972) *Human Information Processing*. Academic Press: New York.

Linthorst, J.M. and B. van Praag (1981) Interaction-patterns and service-areas of local public services in the Netherlands, *Regional Science and Urban Economics*, 11:39-56.

Lloyd, R. and D. Jennings (1978) Shopping behavior and income: comparisons in an urban environment, *Economic Geography*, 54:157-167.

Looney, E. (1987) Using DSI technology for optimizing T1 performance, *Telecommunications*, 21(12):62-67.

Louviere, J.J. and R.J. Meyer (1979) Behavioral analysis of destination choice: theory and empirical evidence, Technical Report 112, Institute of Urban and Regional Research, University of Iowa.

Love R.F. and J.G. Morris (1972) Modelling inter-city road distances by mathematical functions, *Operational Research Quarterly*, 23:61-71.

Love R.F. and J.G. Morris (1979) Mathematical models of road travel distances, *Management Science*, 25:130-139.

Lovett, A.A., I.D. Whyte and K.A. Whyte (1985) Poisson regression analysis and migration fields: the example of the apprenticeship records of Edinburgh in the seventeenth and eighteenth centuries, *Transactions of the Institute of British Geographers*, 10:317-332.

Lowry, I.S. (1966) *Migration and Metropolitan Growth: Two Analytical Models*. Chandler: San Francisco.

Lundqvist, L. (1973) Integrated location-transportation analysis: a decomposition approach, *Regional and Urban Economics*, 3:233-262.

Lussier, D.A. and R.W. Olshavsky (1974) An information processing approach to individual brand choice behaviour. Paper presented at the ORSA/TIMS Joint National Meeting, San Juan, Puerto Rico.

Lycan, R. (1969) Interprovincial migration in Canada: the role of spatial and economic factors, *Canadian Geographer*, 13:237-254.

MacKay, D.B. (1973) A spectral analysis of the frequency of supermarket visits, *Journal of Marketing*, 10:84-90.

Madden, J.F. (1980) Urban land use and growth in two-earner households, *American Economic Review*, 70:191-197.

Manheim, M. (1979) *Fundamentals of Transportation Systems Analysis: Vol 1, Basic Concepts*. M.I.T. Press: Cambridge Mass.

Marble, D.F. (1964) A simple markovian model of trip structure in a metropolitan region, *Papers, Regional Science Association Western Section*, 150-156.

March, L. and M. Batty (1975a) Information minimizing formalism and the derivation of non-parametric forms for population and transportation models, Technical Report 6-S, Dept. of Systems Design, University of Waterloo, Ontario, Canada.

March, L. and M. Batty (1975b) Generalized measures of information, Bayes' likelihood ratio and Jaynes' formalism, *Environment and Planning B*, 2:99-105.

Mayhew, L. and G. Leonardi (1982) Equity efficiency and accessibility in urban and regional health care systems, *Environment and Planning A,* 14:1479-1507.

McCafferty, D., M.E. O'Kelly and M.J. Webber (1979) Disaggregate and aggregate models of destination choice for shopping. Paper presented at the 75th Annual Meetings of the Association of American Geographers, Philadelphia.

McCarthy, P.S. (1980) A study of the importance of generalized attributes in shopping choice behavior, *Environment and Planning A,* 12:1269-1286.

McFadden, D. (1974a) Conditional logit analysis of qualitative choice behavior, in *Frontiers in Econometrics,* P. Zarembka (ed.), Academic Press: New York, pp. 105-142.

McFadden, D. (1974b) The measurement of urban travel demand, *Journal of Public Economics,* 25:303-328.

McFadden, D. (1978) Modelling the choice of residential location, in *Spatial Interaction Theory and Planning Models,* A. Karlquist, L. Lundquist, F. Snickars and J.W. Weibull (eds.), North Holland: Amsterdam, pp. 75-96.

McFadden, D. (1980) Econometric models for probabilistic choice among products, *Journal of Business,* 53:513-529.

McLafferty, S. and A. Ghosh (1986) Multipurpose shopping and the location of retail firms, *Geographical Analysis,* 18(3):215-226.

McLafferty, S. and A. Ghosh (1987) Optimal location-allocation with multipurpose shopping, in *Spatial Analysis and Location-Allocation Models,* A. Ghosh and G. Rushton (eds.), van Nostrand Reinhold: New York, pp. 55-75.

Meyer, R.J. and T.C. Eagle (1982) Context-induced parameter instability in a disaggregate-stochastic model of store choice, *Journal of Marketing Research,* 19:62-71.

Miller, E.J. and M.E. O'Kelly (1983) Estimating shopping destination choice models from travel diary data, *Professional Geographer,* 35:440-449.

Miller, E.J. and M.E. O'Kelly (1985) Reply to Pickles and Davies, *Professional Geographer,* 37:195-196.

Mirchandani, P. and A. Oudjit (1982) Probabilistic demands and costs in facility location problems, *Environment and Planning A,* 14:917-932.

Mood, A. M. and F.A. Graybill (1963) *Introduction to the Theory of Statistics.* McGraw-Hill: New York.

Moore, E.G. and L.A. Brown (1970) Urban acquaintance fields: an evaluation of a spatial model, *Environment and Planning A,* 2:443-454.

Moore, W.L., E.A. Pressemier and D.R. Lehmann (1985) Hierarchical representations of market structures and choice process via preference trees, Working Paper, Department of Marketing, University of Utah.

Mulligan, G. (1983) Consumer demand and multipurpose shopping behavior, *Geographical Analysis*, 15:76-81.

Mulligan, G. (1984) Central place populations: some implications of consumer shopping behavior, *Annals of the Association of American Geographers*, 74:44-56.

Murdie, R.A. (1965) Cultural differences in consumer travel, *Economic Geography*, 41:211-233.

Muth, R.F. (1971) Migration: Chicken or egg? *Southern Economic Journal*, 37:295-306.

Nakanishi, M. and L.G. Cooper (1974) Parameter estimation for a multiplicative competitive interaction model--least squares approach, *Journal of Marketing Research*, 11:303-311.

Narula, S.C., U.I. Ogbu and H.M. Samuelsson (1977) An algorithm for the p-median problem, *Operations Research*, 25:709-713.

Nelder, J.A. and R.W.M. Wedderburn (1972) Generalised linear models, *Journal of the Royal Statistical Society*, Series A 135:370-384.

Newell, A. and H.A. Simon (1972) *Human Problem Solving*. Prentice-Hall: Englewood Cliffs, N.J.

Norman, D.A. and D.G. Bobrow (1975) On data-limited and resource-limited process, *Cognitive Psychology*, 7:44-64.

Nystuen, J. (1967) A theory and simulation of intraurban travel, in *Quantitative Geography, Part 1*, W. Garrison and D. Marble (eds.), Northwestern University Press, Evanston Illinois, pp. 54-83.

O'Brien, L.G. and C.M. Guy (1985) Locational variability in retail grocery prices, *Environment and Planning A*, 17:953-962.

O'Kelly, M.E. (1981a) Generalized information measures, *Environment and Planning A*, 13:681-688.

O'Kelly, M.E. (1981b) A model of the demand for retail facilities incorporating multistop, multipurpose trips, *Geographical Analysis*, 13:134-148.

O'Kelly, M.E. (1981c) *Impacts of multipurpose trip-making on spatial interaction and retail facility size*. Unpublished Ph.D. Thesis, Department of Geography, McMaster University, Hamilton Ontario, Canada.

O'Kelly, M.E. (1983a) Multipurpose shopping trips and the size of retail facilities, *Annals of the Association of American Geographers*, 73:231-239.

O'Kelly, M.E. (1983b) Impacts of multistop, multipurpose trips on retail distributions, *Urban Geography*, 4:173-190.

O'Kelly, M.E. (1986a) The location of interacting hub facilities, *Transportation Science*, 20(2):92-106.

O'Kelly, M.E. (1986b) Activity levels at hub facilities in interacting networks, *Geographical Analysis*, 18(4):343-356.

O'Kelly, M.E. (1987a) Spatial interaction based location-allocation models, in *Spatial Analysis and Location-Allocation Models,* A. Ghosh and G. Rushton (eds.), van Nostrand Reinhold: New York, pp. 302-326.

O'Kelly, M.E. (1987b) A quadratic integer program for the location of interacting hub facilities, *European Journal of Operational Research,* 32:393-404.

O'Kelly, M.E. and E.J. Miller (1984) Characteristics of multistop multipurpose travel: an empirical study of trip length, *Transportation Research Record,* 976:33-39.

O'Kelly, M.E. and J.E. Storbeck (1984) Hierarchical location models with probabilistic allocation, *Regional Studies,* 18(2):121-129.

Olsson, G. (1970) Explanation, prediction and meaning variance: an assessment of distance interaction models, *Economic Geography,* 46:223-233.

Op't Veld, A., E. Bijlsma and J. Starmans (1984) Explanatory analysis of interregional migration in the nineteen-seventies, in *Demographic Research and Spatial Policy,* H. ter Heide and F.J. Willekens (eds.), Academic Press: Orlando, chapter 9.

Openshaw, S. (1973) Insoluble problems in shopping model calibration when the trip pattern is not known, *Regional Studies,* 7:367-371.

Openshaw, S. (1979) Alternative methods of estimating spatial interaction models and their performance in short-term forecasting, in *Exploratory and Explanatory Statistical Analysis of Spatial Data,* C.P.A. Bartels and R.H. Ketellapper (eds.), Martinus Nijhoff: Boston, pp. 201-225.

Oster, C.V. (1978) Household tripmaking to multiple destinations: the overlooked urban travel pattern, *Traffic Quarterly,* 32:511-529.

Ostresh, L. (1973) WEBER in *Computer Programs for Location-Allocation Problems,* G. Rushton, M.F. Goodchild and L. Ostresh (eds.), Department of Geography, Iowa, Mono. No. 6, Iowa City.

Ostresh, L. (1975) An efficient algorithm for solving the two center location-allocation problem, *Journal of Regional Science,* 15:209-216.

Ostresh, L. (1978) On the convergence of a class of iterative methods for solving the Weber location problem, *Operations Research,* 26:597-609.

Papageorgiou, G.J. (1969) On the distribution of certain trip lengths that originate within a given region, Discussion Paper, Department of Geography, Ohio State University.

Papageorgiou, G.J. and A. Brummell (1975) Crude inferences on spatial consumer behavior, *Annals of the Association of American Geographers,* 65:1-12.

Parker (1974) An analysis of retail grocery price variations, *Area,* 6:117-120.

Pas, E. (1982) Analytically derived classification of daily travel-activity behavior: description, evolution and interpretation, *Transportation Research*

Record, 879:9-15.

Payne, C.D. (1985) *The GLIM System Release 3.77.* Numerical Algorithms Group: Oxford.

Peeters, D. (1980) *Contributions aux modeles de localisation de service publics.* Unpublished Ph.D. Thesis, Faculte des Sciences Appliques, Universite Catholique De Louvain.

Peterson, G. (1967) A model of preference: quantitative analysis of the perception of the visual appearance of residential neighborhoods, *Journal of Regional Science,* 7:19-31.

Phillips, F.Y. (1981) *A Guide to MDI Statistics for Planning and Management Model Building.* Institute for Constructive Capitalism: Austin.

Pickles, A.R. (1986) *An Introduction to Likelihood Analysis.* Concepts and Techniques in Modern Geography, 42, Geo Books: Norwich.

Pickles, A.R. and R.B. Davies (1985) Estimating shopping destination choice models from travel diary data: a comment on omitted variables, *Professional Geographer,* 37:194-195.

Pipkin, J.S. (1978) Fuzzy sets and spatial choice, *Annals of the Association of American Geographers,* 68:196-204.

Pipkin, J.S. and D.P Ballou (1979) A model of central city and suburban trip termination patterns, *Journal of Regional Science,* 19(2):179-190.

Pitfield, D.E. (1978) Sub-optimality in freight distribution, *Transportation Research,* 12:403-409.

Plane, D.A. (1984) Migration space: doubly-constrained gravity model mapping of relative interstate separation, *Annals of the Association of American Geographers,* 74:244-256.

Potter, R.B. (1977) Spatial patterns of consumer behaviour and perception in relation to the social class variable, *Area,* 9:153-156.

Powell, M.J.D. (1970) A FORTRAN subroutine for solving systems of non-linear algebraic equations, in *Numerical Methods for Non-linear Algebraic Equations,* P. Rabinowitz (ed.), Gordon and Breach: London.

Preston, V. and S. Takahashi (1983) Multistop trips: some empirical findings. Paper presented at the Annual Meetings of the Association of American Geographers, Denver, CO, April 1983.

Preston, V. and S.M. Taylor (1981) Personal construct theory and residential choice, *Annals of the Association of American Geographers,* 71:437-451.

Rao, V.R. and D.J. Sabavala (1981) Inference of hierarchical choice processes from panel data, *Journal of Consumer Research,* 8:85-96.

Ravenstein, E.G. (1876) The birthplaces of the people and the laws of migration, *The Geographical Magazine,* 111:173-177, 201-206, 229-233.

Ravenstein, E.G. (1885) The laws of migration, *Journal of the Royal Statistical Society*, 48:167-235.

Ravenstein, E.G. (1889) The laws of migration: second paper, *Journal of the Royal Statistical Society*, 52:241-305.

Ravindran, A. (1972) A computer routine for quadratic and linear programming problems, Algorithm 431, *Communications of the ACM*, 15:818-820.

Ravindran, A. and H.K. Lee (1981) Computer experiments on quadratic programming algorithms, *European Journal of Operational Research*, 8:166-174.

Ray (1967) Cultural differences in consumer travel behavior in Eastern Ontario, *Canadian Geographer*, 11:143-156.

Recker, W. and L. Kostyniuk (1978) Factors influencing destination choice for urban grocery shopping trips, *Transportation*, 7:19-33.

Reilly, W.J. (1931) *The Law of Retail Gravitation*. Putman: New York.

Restle, F. (1961) *Psychology of Judgement and Choice: A Theoretical Essay*. Wiley: New York.

ReVelle, C. and R. Swain (1970) Central facilities location, *Geographical Analysis*, 2:30-42.

Richardson, A. (1982) Search models and choice set generation, *Transportation Research*, 16A:403-420.

Robinson, R.V.F. and R.W. Vickerman (1976) The demand for shopping travel: a theoretical and empirical study, *Applied Economics*, 8:267-281.

Rogers, A. (1965) A stochastic analysis of the spatial clustering of retail establishments, *Journal of the American Statistical Association*, 60:1094-1103.

Rogers, A. (1967) A regression analysis of interregional migration in California, *The Review of Economics and Statistics*, 49:262-267.

Ross, G.T. and R.M. Soland (1977) Modeling facility location problems as generalized assignment problem, *Management Science*, 24(3):345-357.

Rossi, P.H. (1955) *Why Families Move: A Study in the Social Psychology of Urban Residential Mobility*. Free Press: Glencoe, Illinois.

Ruefli, T.W. and J.E. Storbeck (1984) Behaviorally-linked location hierarchies, *Environment and Planning B*, 9:257-268.

Sasaki, T. (1972) Estimation of person trip patterns through markov chains, in *Theory of Traffic Flow and Transportation: Proceedings of the Fifth International Symposium*, G.F. Newell (ed.), Elsevier: New York, pp. 119-130.

Schuler, H.J. (1979) A disaggregate store-choice model of spatial decision-making, *Professional Geographer*, 31:146-156.

Schwartz, A. (1973) Interpreting the effect of distance on migration, *Journal of Political Economy*, 81:1153-1169.

Schwartz, M. (1987) *Telecommunication Networks: Protocols, Modeling, and Analysis.* Addison-Wesley: Reading Mass.

Scott, A. (1971) A model of nodal entropy in a transportation network with congestion costs, *Transportation Science*, 5:204-211.

Scott, P. (1970) *Geography and Retailing.* Hutchinson Univ. Library: London.

Sen, A. and R.K. Pruthi (1983) Least squares calibration of the gravity model when intrazonal flows are unknown, *Environment and Planning A*, 15:1545-1550.

Sen, A. and S. Soot (1981) Selected procedures for calibrating the generalized gravity model, *Papers of the Regional Science Association*, 48:165-176.

Senior, M.L. (1979) From gravity modelling to entropy maximising: a pedagogic guide, *Progress in Human Geography*, 3:174-210.

Shannon, C.F. (1948) A mathematical theory of communication, *Bell System Technical Journal*, 27:379-423 and 623-656.

Shepard, I.D. and C.J. Thomas (1980) Urban consumer behaviour, in *Retail Geography*, J.A. Dawson (ed.), Croom Helm: London.

Sheppard, E.S. (1978) Theoretical underpinnings of the gravity hypothesis, *Geographical Analysis*, 10:386-402.

Sheppard, E.S. (1980) Location and the demand for travel, *Geographical Analysis*, 12:111-128.

Sheridan, J.E., M.D. Richards and J.W. Slocum (1975) Comparative analysis of expectancy and heuristic models of decision behaviour, *Journal of Applied Psychology*, 60:361-368.

Simon, H.A. (1969) *The Science of the Artificial.* MIT Press: Cambridge, MA.

Smith, G.C. (1976) The spatial information fields of urban consumers, *Transactions of the Institute of British Geographers*, New Series 1:175-189.

Snickars, F. (1978) Convexity and duality properties of a quadratic intraregional location model, *Regional Science and Urban Economics*, 8:5-20.

Snickars, F. and J.W. Weibull (1977) A minimum information principle: theory and practice, *Regional Science and Urban Economics*, 7:137-168.

Southworth, F. (1983) Temporal versus other impacts upon trip distribution model parameter values, *Regional Studies*, 17:41-47.

Stetzer, F. (1976) Parameter estimation for the constrained gravity model: a comparison of six methods, *Environment and Planning A*, 8:673-683.

Stillwell, J.C.H. (1978) Interzonal migration: some historical tests of spatial-interaction models, *Environment and Planning A*, 10:1187-1200.

Stopher, P.R. and A.H. Meyburg (1979) *Survey Sampling and Multivariate Analysis for Social Scientists and Engineers*. Heath: Lexington, MA.

Stronge, W.B. and R.R. Schultz (1978) Heteroscedasticity and the gravity model, *Geographical Analysis*, 10:279-286.

Swait, J. and M. Ben-Akiva (1987a) Incorporating random constraints in discrete models of choice set generation, *Transportation Research*, 21B:91-102.

Swait, J. and M. Ben-Akiva (1987b) Empirical test of a constrained choice discrete model: mode choice in Sao Paulo, Brazil, *Transportation Research*, 21B:103-115.

Taaffe, E.J., R.L. Morrill and P.R. Gould (1963) Transportation expansion in underdeveloped countries: a comparative analysis, *Geographical Review*, 53:503-529.

Takahashi, S. (1986) A case study of the transferability of a retail choice model. Paper presented at the Annual Meetings of the Association of American Geographers, Minneapolis, MN, April 1986.

Tapiero, C.S. (1974) The demand and utilization of recreational facilities: a probability model, *Regional and Urban Economics*, 4(2):173-185.

Teitz, M.P. and P. Bart (1968) Heuristic methods for estimating the generalized vertex median of a weighted graph, *Operations Research*, 16:955-961.

Thill, J.-C. and I. Thomas (1987) Toward conceptualizing trip-chaining behavior: a review, *Geographical Analysis*, 19(1):1-17.

Thomas, C.J. (1974) The effects of social class and car ownership on intra-urban shopping behaviour in Greater Swansea, *Cambria*, 2:98-126.

Thomas, R.W. (1977) An interpretation of the journey to work on Merseyside using entropy-maximizing methods, *Environment and Planning A*, 9:817-834.

Thompson, B. (1967) An analysis of supermarket shopping habits in Worcester, Massachusetts, *Journal of Retailing*, 43:17-29.

Timmermans, H.J.P. and K.J. Veldhuisen (1981) Behavioural models and spatial planning: some methodological considerations and empirical tests, *Environment and Planning A*, 13:1485-1498.

Tobler, W. (1979) Estimation of attractivities from interactions, *Environment and Planning A*, 11:121-127.

Tobler, W. (1983) An alternative formulation for spatial-interaction modeling, *Environment and Planning A*, 15:693-703.

Tomlin, J.A. and S.G. Tomlin (1968) Traffic distribution and entropy, *Nature*, 220:974-976.

Toyne, P. (1971) Customer trips to retail business, in *Exeter Essays in Geography*, W.L.D. Ravenhill and K.J. Gregory (eds.), Exeter Univ. Press: Exeter.

Train, K. (1986) *Qualitative Choice Analysis.* MIT Press: Cambridge, MA.

Trew, R.D. (1988) *Supermarket choice behavior: a multinomial logit model of grocery shopping patterns in Gainesville, Florida.* Unpublished M.A. Thesis, Dept. of Geography, University of Florida, Gainesville.

Tribus, M. (1969) *Rational Descriptions Decisions and Designs.* Pergamon Press: New York.

Tribus, M. and R. Rossi (1973) On the Kullback information measure as a basis for information theory: comments on a proposal by Hobson and Cheng, *Journal of Statistical Physics,* 9:331-338.

Tversky, A. (1972) Elimination by aspects: a theory of consumer choice, *Psychological Review,* 79:281-299.

Tversky, A. and S. Sattath (1979) Preference trees, *Psychological Review,* 86:542-593.

Vanderkamp, J. (1971) Migration flows: their determinants and the effects of return migration, *Journal of Political Economy,* 79:1012-1031.

van Est, J. and J. van Setten (1979) Two estimation methods for singly constrained spatial distribution models, in *Exploratory and Explanatory Statistical Analysis of Spatial Data,* C.P.A. Bartels and R.H. Ketellaper (eds.), Martinus Nijhoff: Boston, pp. 227-241.

van Lierop, W. and P. Nijkamp (1982) Disaggregate models of choice in a spatial context, *Sistemi-Urbani,* 413:331-369.

Walsh, J.A. (1976) Information theory and classification in geography, DP-4, Department of Geography, McMaster University, Hamilton, Ontario.

Walsh, J.A. (1980) An entropy maximising analysis of journey-to-work patterns in county Limerick, *Irish Geography,* 13:33-53.

Walsh, J.A. and M.J. Webber (1977) Information theory: some concepts and measures, *Environment and Planning A,* 9:395-417.

Weaver, J.R. and R.L. Church (1985) A median facility location model with nonclosest facility service, *Transportation Science,* 19:58-74.

Webber, M.J. (1975) Entropy-maximising location models for nonindependent events, *Environment and Planning A,* 7:99-108.

Webber, M.J. (1977) Pedagogy again: what is entropy? *Annals of the Association of American Geographers,* 67(2):254-266.

Webber, M.J. (1978) Spatial interaction and the form of the city, in *Spatial Interaction Theory and Planning Models,* A. Karlquist, L. Lundquist, F. Snickars and J.W. Weibull (eds.), North Holland: Amsterdam, pp. 203-226.

Webber, M.J. (1979) *Information Theory and Urban Spatial Structure.* Croom Helm: London.

Webber, M.J. (1980) Urban spatial structure: and information theoretic approach, unpublished report to SSHRC, 410-77-0582, Department of Geography, McMaster University, Hamilton, Ontario.

Webber, M.J. and M.E. O'Kelly (1981) Empirical tests and sensitivity analysis of a model of residential and facility location, *Geographical Analysis*, 13(4):398-411.

Webber, M.J., M.E. O'Kelly and P.D. Hall (1979) Empirical test on an information minimising model of consumer characteristics and facility location, *Ontario Geography*, 13:61-80.

Wesolowsky, G. and R.F. Love (1972) A nonlinear approximation method for solving a generalized rectangular distance Weber problem, *Management Science*, 18:656-663.

West, D.H. (1983) Approximate solution of the quadratic assignment problem, Algorithm 608, *Transactions on Mathematical Software*, 9:461-466.

Wheeler, J.O. (1972) Trip purposes and urban activities linkages, *Annals of the Association of American Geographers*, 62:641-654.

Williams, H.C.W.L. (1977) On the formation of travel demand models and economic evaluation measures of user benefit, *Environment and Planning A*, 9:285-344.

Williams, H.C.W.L. and J.D. Ortuzar (1982) Behavioral theories of dispersion and the mis-specification of travel demand models, *Transportation Research B*, 16:167-219.

Williams, H.C.W.L. and M.L. Senior (1977) A retail location model with overlapping market areas: Hotelling's problem revisited, *Urban Studies*, 14:203-205.

Williams, P. (1988) A recursive model of intraurban trip-making, *Environment and Planning A*, forthcoming.

Williams, P.A. and A.S. Fotheringham (1984) *The Calibration of Spatial Interaction Models by Maximum Likelihood Estimation with Program SIMODEL*. Geographic Monograph Series 7, Department of Geography, Indiana University.

Willmott, C.T. (1984) On the evaluation of model performance in physical geography, in *Spatial Statistics and Models*, G.L. Gaile and C.T. Willmott (eds.), Reidel: New York.

Wilson, A.G. (1967) Statistical theory of spatial trip distribution models, *Transportation Research*, 1:253-269.

Wilson, A.G. (1969) Generalizing the Lowry model, CES-WP56, Centre for Environmental Studies, London.

Wilson, A.G. (1974) *Urban and Regional Models in Geography and Planning*. Wiley: London.

Wilson, A.G. (1975) Some new forms of spatial interaction model: a review, *Transportation Research*, 9:167-179.

Wilson, A.G. (1981a) *Catastrophe Theory and Bifurcation: Applications in Urban and Regional Geography*. Croom Helm: London.

Wilson, A.G. (1981b) The evolution of urban spatial structure: the evolution of theory, in *European Progress in Spatial Analysis*, R.J. Bennett (ed.), Pion: London, pp. 201-225.

Wilson, A.G., J.D. Coehlo, S.M. Macgill and H.C.W.L. Williams (1981) *Optimization in Locational and Transport Analysis*. Wiley: London.

Wilson, A.G. and M.L. Senior (1974) Some relationships between entropy maximizing models, mathematical programming models and their duals, *Journal of Regional Science*, 14:207-215.

Wrigley, N. (1985) *Categorical Data Analysis*. Longmans: London.

Wrigley, N. and R. Dunn (1984) Stochastic panel-data models of urban shopping behaviour: 1. Purchasing at individual stores in a single city, *Environment and Planning A*, 16:629-650.

Zadeh, L.A. (1965) Fuzzy sets, *Information and Control*, 8:338-353.

# Subject Index